# CORN
# MEALS & MORE

## By Olwen Woodier

A Garden Way Publishing Book

STOREY

Storey Communications, Inc.
Schoolhouse Road
Pownal, VT 05261

## ACKNOWLEDGMENTS

A big thanks to:

My editor Jill Mason who helped pull this book together

My daughter Wendy for eating lots of muffins and cornbreads

My husband Richard Busch for getting me through a tough time

Dr. William Pardee, Chairman of Plant Breeding at Cornell University, for his insight into crossbreeding and the different types of corn

and finally, thanks to the wonderful people at Katonah Library who helped me locate reference and research materials, and anyone else who gave me information or recipes on corn.

*Cover Illustration by Chris Spollen*

*Cover Design by Leslie Morris Noyes*

*Text Design by Cindy McFarland*

*Text Illustrations by Sue Storey*

Printed in the United States by Banta Company

First Printing, May 1987

**Library of Congress Cataloging-in-Publication Data**

Woodier, Olwen, 1942–
    Corn: meals and more.
    "A Garden Way Publishing book."
    Includes index.
    1. Cookery (Corn)   2. Corn.   I. Title.
TX809.M2W66   1987        641.6'315        86-45972
ISBN 0-88266-456-5

# CONTENTS

*This book is for Wendy,*
*the daughter of a British immigrant mother*
*who recently pledged allegiance to the Stars and Stripes,*
*and a Yankee father*
*whose ancestors were growing corn in Massachusetts*
*during the seventeenth century.*

# CORN

## AN INTRODUCTION

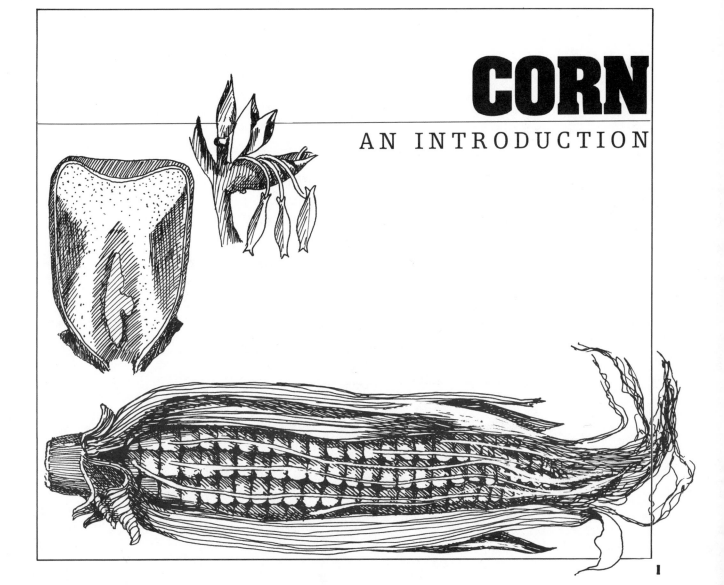

*The rose may bloom for England,*
*The lily for France unfold;*
*Ireland may honor the shamrock,*
*Scotland her thistle bold;*
*But the shield of the great Republic,*
*The glory of the West,*
*Shall bear a stalk of the tasselled corn—*
*The sun's supreme bequest!*

—EDNA DEAN PROCTOR (1838–1923)
COLUMBIA'S EMBLEM

I am not one to wish the days away, but whenever I drive along country roads and pass fields that are green with the promise of corn, I feel an overwhelming impatience with the month of June and wish that it were already mid-July. By then the tall field sentinels have finally tasseled and "Fresh Corn" signs adorn all the roadside vegetable stands. Surely there is nothing more sublime than that first fresh cob of the season—or more disappointing if it was picked too young or exposed to heat so that the sweetest of milk sugars either has not developed or has already turned to the starch.

Eating fresh corn on the cob is a gastronomic experience that defies all understanding of etiquette. Forget the special corn holders and nibbling daintily along the kernels. The only way to eat corn—if you are a true corn lover—is to take it steaming hot in hand, slather it with butter, lightly sprinkle it with salt, and devour two or more rows at a time. This is not a feast to offer the boss, or anyone else who needs impressing. Only family and close friends will tolerate the slurping and messiness and the contented groans as the last ear is consumed.

## THE HISTORY OF CORN

The great civilizations of the Incas, Mayas, and Aztecs were founded on the cultivation of corn. Indeed, agriculture was the very center of their religion. Corn was their life source and they depended on the rain to make it grow. In the name of corn, the Aztec people even sacrificed young women to two of their important deities, the rain god Tlaloc and the corn goddess Chicomecoatl.

Corn was also the staple grain of the North American Indians. Many sacred rituals were developed to summon the rain to ensure a good harvest. Some of these rain ceremonies, like the Pueblo Indians' Hopi Serpent Dance, are still practiced to encourage the growth of corn struggling in the parched earth. As part of the Hopi Serpent Dances rattlesnakes, sprinkled with sacred cornmeal, are carried into the desert and released. This practice is believed to have originated with a prehistoric Mexican rattlesnake cult.

Native American Indians called corn *mahiz*, which meant "our life" in their language (other tribes called it *carracony*, *pegatowr*, and *roanoke*).

Until 1950 corn was thought to have been brought to the New World from Asia during prehistoric migrations. But in that year anthropologists discovered, 200 feet below Mexico City, fossilized grains of pollen believed to be the pollen of wild corn. By carbon-dating methods, this pollen was determined to be eighty thousand years old, predating the arrival of human beings in this hemisphere and proving that corn was indeed indigenous to the North American continent. Between 1945 and 1960, other expeditions unearthed tiny ears of *cultivated* corn, half an inch long, that were carbon-dated to 4000–3000 B.C. The most notable finds were made at Tehuacán Valley in Mexico by Canadian anthropologist Dr. Richard MacNeish, and at Bat Cave in New Mexico by Dr. Herbert Dick of Harvard University's Peabody Museum.

## THE CORN WAR

The evolution of domestic corn remains something of a controversy. Because the husks firmly contain the kernels, domestic corn cannot reproduce successfully without humans to assist in freeing the seeds, or kernels, from the cob and planting them in the ground. Corn kernels are too heavy to float on the wind; nor are they carried away by birds, since birds eat them the way they eat berries, on the spot. Until recently, scientists were unable to determine how wild corn had propagated without human intervention. In 1961, however, Dr. MacNeish discovered, near Tehuacán, tiny 7,200-year-old cobs of corn still bearing a few kernels. In these ancient cobs the ears were enclosed in a husk that opened when the corn was ripe, allowing the small, light kernels to be dispersed by the wind. As it cross-pollinated and developed the characteristics of cultivated corn, wild corn could no longer release its husk and kernels. Eventually, wild corn became extinct.

A common theory during the first half of this century held that domesticated corn was developed from wild teosinte grasses. Corn geneticist Paul Mangelsdorf, however, declared the half-inch cobs of wild corn to be the true ancestors of modern-day corn instead. His rejection of the widely held teosinte theory triggered an intellectual corn war within the scientific establishment. Mangelsdorf's leading opponent was geneticist George W. Beadle, a staunch believer that teosinte was the ancestral grain, and scholarly papers were furiously penned between the two camps for more than thirty years.

Then in 1977 Mexican botanist Rafael

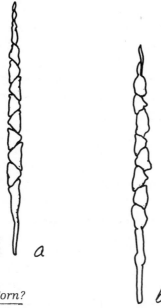

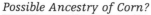

*Possible Ancestry of Corn?*

*Geneticist George W. Beadle
has reconstructed the domestication
of corn from the wild grass teosinte over
thousands of years. The illustrations on this page are
approximately life-size. In (a) the teosinte "spike" (similar to
the corn "ear") bears a single row of hard-shelled seeds, or kernels,
which scatter when ripe. Illustrations (b) and (c) show possible transitional stages.
In (d), a primitive, 2-inch ear of corn, similar to the 7,000-year-old ears unearthed in Tehuacán,
was obtained by crossing modern corn with teosinte. Modern corn when ripe, does not release its seeds,
and thus cannot survive without human assistance.*

Guzman discovered wild perennial teosinte growing in a five-acre area near Jalisco, in southwestern Mexico. Like modern corn, it had twenty chromosomes and thus was proven to be its most primitive known relative.

In response to this apparent defeat, Dr. Mangelsdorf wrote: "The long debated question, 'Which is the ancestor of cultivated corn—teosinte or wild corn?' is no longer relevant. Both are."

## CORN TRAVELS NORTH

It probably took centuries for corn to be carried north from Mexico by Indians migrating to the Four Corners region of New Mexico and Arizona, and still centuries more to reach native Americans living in the Northeast. In the fifteenth century, the Spaniards discovered that corn was grown by tribes all the way from the tip of South America to Canada. Returning from a post in Cuba in 1492, a scouting party, informed its leader, Christopher Columbus, that this "mahiz" could be baked, dried, and made into flour. Within a few years, *maize* had been introduced to Europe. The Portuguese took maize seeds with them on their travels to Africa, the East Indies, and Asia, and by the late sixteenth century, corn was being grown in many corners of the world.

The British colonists in Jamestown, Virginia, would have starved to death in 1607 if they hadn't planted corn under the guidance of Captain John Smith, who had procured the seeds from the Powhatan tribe. Hoping to find gold when they arrived, Smith's colony finally found it in the corn they grew. The people of Jamestown were so successful with their corn harvests that by 1630, they were exporting corn to Massachusetts and the Caribbean. During the late seventeenth century, when legal currency was insufficient to meet public demand, the Corn Exchange Bank was established and corn kernels were used in place of money.

The Indians' method of preserving maize under mounds of sand also saved the Pilgrims in 1620, during their first winter in Massachusetts. When the Mayflower arrived at Plymouth, the Pilgrims were almost without supplies. Walking inland, they came across a buried stash of corn, which not only fed them during that winter, but also left them with enough to plant in the spring.

These settlers found that corn grew to maturity within several weeks of planting and yielded four times more per acre than their own British grains (wheat, for example, takes ten months from planting to harvest). They called it "Indian corn" because of its origins and the fact that all grains (wheat, barley, oats, rye) are lumped together as "corn" in Anglo-Saxon Britain. From the Indians, they learned to use it in a variety of ways and, experimenting with the native methods for making cornmeal breads, developed their own recipes. The results were vastly different from crusty well-risen wheat breads.

John Josselyn, a chronicler of the time, wrote about cornmeal mush and corn bread, calling them "New England standing Dishes." During the 1670s, he observed that

*"Indian Wheat, of which there is three sorts, yellow, red, and blew . . . is light of digestion, and the English make a kind of Loblolly of it to eat with Milk, which they call Sampe; they beat it in a Mortar, and sift the flower out of it; the remainder they call Homminey, which they put into a Pot of two or three Gallons with water and boyl it upon a gentle Fire till it be like a Hasty Pudden; they put of this into Milk, and eat it. Their bread also they make of Homminey so boyled, and mixe their Flower with it. . . ."*

It wasn't until 1779 that the settlers had their first taste of sweet corn (up to that time they had been growing field corn). First grown in the Susquehanna River Valley by the Iroquois, it was brought to New England, Pennsylvania, and eastern New York by the military, who had pillaged villages of the Indians loyal to Britain. Called *popoon* by the Iroquois nations, the sweet corn had juicy white kernels clinging to five-inch crimson cobs.

During the eighteenth century, Americans began to diversify their agriculture. The grain the Indians and early settlers had consumed in all its forms—fresh, dried, parched, and ground—was slowly edged aside in colonial kitchens to make room for wheat, oat, and potato dishes.

## CORN BELT

Although corn lost some of its importance as a staple food on North American tables, it continued to grow as a major industry. During the first half of the nineteenth century, Virginia, North Carolina, Kentucky, and Tennessee were the main corn-growing states. But as the railroad pushed into the

Corn has been the subject of paintings, sketches, cartoons, books, almanacs, poems, and songs. Writers, in fact, have been waxing lyrical about America's golden grain for at least two hundred years. As you might imagine, most of the facts and fiction have been penned by Americans, at home and abroad. The children of America grew up singing verses from John Greenleaf Whittier's *The Corn Song* or quoting passages like:

*All around the happy village*
*Stood the maize fields, green and*
*    shining,*
*Waved the green plumes of*
*    Mondamin,*
*Waved his soft and sunny tresses,*
*Filling all the land with plenty.*

from the *Song of Hiawatha* by Henry Wadsworth Longfellow. (Indeed, *Hiawatha* was a universal favorite. Given to me by my father during my early teen years in England, it was a book that I loved to curl up with and dream not so much of corn but of being an Indian maiden, dressed in buckskin and riding bareback.)

It's obvious from the number of corn jokes that found their way into the American mainstream that corn has been the focus of much humor. The expression "corny" developed from "corn-fed," a disparaging remark that theatrical groups made about their unsophisticated audiences from small towns in corn-growing country. If

you were fat, you could also be called cornfed. Originally, corny meant out-of-date or old-fashioned music—not so different from the way we use it today.

Sprinkled throughout the American language, the word *corn* is used in many expressions, some of which are directly related to food. "Corned," for example, refers to a method of preserving meats and fish in brine. The original system was to layer the flesh with granular salt, sugar, and saltpeter (which retains the color), and because corn was used to describe any small particle as well as a small grain, the whole process was called "corning." When you ate corned beef in those days, you were eating "corn willie." If you were in the Army during World War II, the dish you know as corned beef hash at home would be served up as "corn bill." "Corn cracker" was used to described a good thing, and "corn in Egypt" indicated there was plenty of food.

To call someone just plain "corned" meant that he was pickled or drunk, but if he was "corny-faced," his face was red and pimply from drink. If you drank "corn juice," you were imbibing inferior or homemade whiskey, and when you brought out the "corn mule," you were serving not only homemade but bootleg whiskey.

And, for sports fans, a "can of corn" is an easy play in baseball.

Who knows, there may be a kernel of truth in all of this.

Midwest, corn growing exploded and Illinois, Iowa, Indiana, Missouri, Kansas, Ohio, Nebraska, Minnesota, Michigan, Wisconsin, and North and South Dakota became the big corn producers, collectively known as the Corn Belt.

By the late 1800s, the corn industry was booming and played such a role in the economy of this country that the golden grain was referred to as King Corn and Queen Corn. To honor this royal wonder, giant edifices were erected in Iowa, Kansas, and South Dakota. The cob-covered Corn Palace in Mitchell, South Dakota, was constructed in 1892 and still attracts flocks of tourists. Sheathed in corn cobs, it takes the staggering quantity of three thousand bushels of corn to reshingle it each year.

At the turn of the twentieth century, children in Illinois were encouraged to observe an annual Corn Day and were asked to write essays on "Why Corn Should Be Our Emblem." Corn clubs abounded for young and old alike, and corn shows and exhibitions were popular events. The Illinois College of Agriculture even conducted classes to train professional judges of corn, and in 1902 a *Manual of Corn Judging* was published. It outlined the ideal shape, length, number of rows, color, and weight of a winning ear of corn. Each corn state set its own criteria for a perfect ear. Corn was serious business everywhere.

## BIG BUSINESS

Representing more than $20-billion a year, corn is America's biggest money-making crop. It is grown on twenty-five percent of this country's crop land, and over seven billion bushels of corn have been harvested each year since 1978.

A crop that tolerates a wide range of temperatures and soil variations, corn can be grown from the tropics to the Canadian border. However, over 50 percent of the United States' corn crop is produced in the fertile flat region of the Corn Belt, where the climate is favorable and the soil is friable and well drained.

Although it might seem to Americans that we eat lots of corn on the cob, in actual fact only one percent of what is grown here is eaten here—fresh, canned, or frozen. Over fifty percent of what is grown in the Corn Belt states is used as fodder to fatten livestock—it's not only fed to beef and dairy cattle, but it is also used to fatten hogs and poultry. (Yellow-colored chickens are fed yellow corn and white-fleshed chickens are given white corn.) Twenty-five percent is

## CORN HUSKS AND STALKS

Husks were wrapped around flat cakes before they were baked in ashes, and they were rolled skin-like around sausage meat. This versatile papery material was also transformed into writing paper and cigarette wrappers—which were filled with dried corn-silk tobacco. Shredded husks were stuffed into cushions, pillows, and mattresses, and clumps were bound to a stick and fashioned into brooms to sweep the floors and furniture. Their uses were not all utilitarian. They were made into decorative dry flowers and cornhusk dolls, and were braided for weaving into mats and chair covers.

Cornstalks didn't go to waste either. Children chewed on them for their sugary sweetness, and the elders crushed them for syrup. Rows of stalks were used for fencing and roofing construction. Bundles were packed up against cabin walls as insulation. Inside the house, they were used as floor coverings and for screening off sleeping quarters.

At the end of the corn season, farm animals were turned loose in fields to eat the stalks, leaves, and green ears of corn. The stalks were also chopped and soaked in hot water, or mixed with molasses (which was extracted from the corn cobs) and cornmeal for animal feed. Corn grain, whole or crushed, has been fed to horses, sheep, hogs, cattle, and poultry since the days of the early settlers, who recognized that it was the best feed for fattening up their livestock.

exported to countries around the globe, and the rest is converted into a myriad of by-products. The obvious ones include breakfast cereal, cornmeal, grits, cornstarch, corn oil, corn syrup, and popcorn. We also consume corn indirectly when we drink milk and when we eat steak, pork, chicken, sausages, cheese, butter, and other livestock products.

To derive a number of corn by-products, processors separate the kernel into four components: starch, germ, fiber, and protein. The starch, for example, is converted into corn syrups, dextrose, and high-fructose corn syrup. Liquid dextrose is fermented to produce ethanol (for lead-free automotive fuels) and carbon dioxide (used in hydroponic greenhouses, carbonated beverages, and refrigeration equipment). The germ goes into cooking oils, margarine, mayonnaise, salad dressings, and shortening; the hulls (fiber) and protein produce gluten feed and gluten meal; soluble portions are also used in foodstuffs.

Corn by-products are used for coloring, binding, thickening, coating, and sweetening. Thousands of products found on the supermarket shelves rely on corn as an ingredient, including gelatin, ice cream, wine, fruit drinks, soft drinks, jams, peanut butter, vinegar, chewing gum, bakery products, baby foods, marshmallows, breakfast foods, soups, gravy, olives, processed cheese, vitamin preparations, and medicines. In one form or another, corn is used in wallboard, packaging, oil drilling, plastics, paper, insulation, soap, chemicals, adhesives, printing, solvents, batteries, gum, tobacco, denim jeans, and cosmetics.

## TYPES OF CORN

Corn has been a mainstay of the American diet ever since the Indians put a hoe to the ground. The first type of corn was the type we know as popcorn. By selecting the largest and finest kernels of popcorn for cultivation, the Indians gradually developed more types of corn—flint (Indian), dent (field), and sweet. To this day, there are four major types of corn. Only through recent hybridization have the geneticists developed subclasses, including a high-lysine field corn (which contains a more complete protein than other corn types), floury corn (a softer corn for grinding, used mainly by the Seneca Indians), and super sweets, the modern versions of the Indians' sweet corn. Bred for specific purposes, the four major types of corn differ in a number of ways.

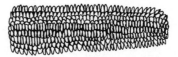

*Popcorn*

**Popcorn** (*Zea mays everta*): The ears are short and the kernels have a hard outer shell surrounding a soft core of starch. When heated, the kernels explode and the moist center bursts out of the hull. Some varieties —Bloody Butcher and Strawberry—are almost too pretty to pop, and many people use them for Thanksgiving and Christmas decorations.

*Flint (Indian) corn*

**Flint or Indian Corn** (*Zea mays indurata*): This corn got its name from the hard, smooth, flint-like kernel. It is usually sold as decorative corn and is a favorite in the fall at roadside stands. It comes in blue, red, black, purple, orange, and multicolored. Don't be misled by the colors: it is edible when ground and makes a sweet, high-protein cornmeal for breads, muffins, and pancakes.

*Dent (Field) corn*

**Dent or Field Corn** (*Zea mays indenta*): The ears are large and the kernels, which have a soft, starchy center, develop a dent when dried. This is *the* commercial corn. It is grown for animal feed, ethylene alcohol, breakfast flakes, corn syrup, and corn starch; this corn grinds finer than the harder kernels of flint. Although you'd never know, some five hundred products contain it in some form. When it is picked as soon as it ripens, it is sweet enough for human consumption, but it has a tough hull.

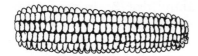

*Sweet corn*

**Sweet Corn** (*Zea mays rugosa*): This is the one for eating fresh, frozen, or canned. It stores its carbohydrates as sugar, which changes to starch as the corn ages. Picking the corn speeds up the conversion of sugar to starch (as does removing the husk). However, today many sweet and super-sweet varieties are still sweet after several hours, and the breeders now aim for those that become even sweeter after harvesting.

## NATIVE AMERICAN CORN BREEDING

Although three hundred varieties of corn were already in existence at the time of the Incan culture, the American Indians continued to produce and cross-breed their own varieties. In order to maintain distinct strains and colors, the Indians planted different varieties separately to avoid contamination through windblown pollen. Some Indians, particularly the Senecas, crossed the corn in one village with a variety from another village. This practice was called "marrying" the corn. The Mandans of Missouri were especially successful farmer-breeders and grew several varieties, including soft and hard white, yellow, red, spotted, blue, pink, black, streaked red, and sweet. The variation and adaptability in their corn strains enabled these native farmers to grow corn in desert, temperate, and tropical environments.

Three parts of the kernel influence the color. The endosperm (the "germ" that nourishes the embryo) is made up of white, yellow, or orange-yellow pigments. The second layer of aleurone and the outer hull (the pericarp) may be blue, yellow, red, brown, or even colorless. Because the aleurone and pericarp are translucent colors, the deep color of the endosperm shows through. The color we see is actually a combination of the three colored layers.

The six main colors of corn—black, red, white, yellow, blue, and multicolored, also a result of the Indians' methods of propagation—were chosen for spiritual interpretations. (According to Pueblo lore, black corn possessed dangerous powers that only medicine men could control.) The six colors repsented north, east, south, west, zenith, and nadir, although the meanings of the colors would vary among different tribes.

The English settlers inherited the basic types of corn from the Atlantic Coast Indians and developed a number of their own shades and varieties. Because of their high yield, the most popular varieties were yellow and white of the dent and flint types.

## HYBRIDIZATION

Up to the early 1900s, the development of colors and high-yield corns was achieved through the simple technique of planting two varieties close enough together to cross pollinate. Each was pollinated by the airborne pollen of the other. The Indians and settlers also practiced breeding sweet corn, selecting specific characteristics such as super sweetness, vigor, and early maturation. Eventually, this led to the development of modern hybridization—cross-breeding from parent plants.

Quiet rumblings were going on in the hybrid field during the first decade of the twentieth century, but it wasn't until 1918 that successful cross-breeding was achieved. By 1920, Henry Agard Wallace (Franklin Roosevelt's third-term vice president and a corn-growing, corn-husking, Iowa native) jumped on the bandwagon and contributed to "the harnessing of hybrid power"—the controlled cross-breeding of different corn varieties to produce the most desirable and vigorous traits of the parent plants. The first hybrid, called Red Green, was released by the Connecticut Agricultural Experiment Station in 1924.

Wallace opened his own company, "The Pioneer Hi-Bred Corn Company," in Des Moines, Iowa. After a ten-year struggle, it took off gangbuster style when his corn was unaffected by the great windstorm of 1936. Pioneer's sales that year amounted to six million kernels.

*And he gave it for his opinion, that whoever could make two ears of corn to grow upon a spot of ground where only one grew before, would do more essential service to his country, than the whole race of politicians put together.*

—JONATHAN SWIFT
GULLIVER'S TRAVELS

The year 1950 saw a breakthrough for sweet corn. Dr. J. R. Laughnam at the University of Illinois discovered a corn cross with more sugar than standard sweet corn. Because the dried kernels had a shriveled appearance, they were named "shrunken two," or SH2, and marketed as "super sweet," "ultra sweet," and "extra sweet." In the 1960s, an even sweeter and more tender corn was produced at the University of Illinois. Dr. A. M. Rhodes discovered a three-way cross and called it sugary enhanced (SE). (The seed catalogs list this corn as Everlasting Heritage, or EH). The sugar levels in these two genetic types of sweet corn produce ears that retain their sweet flavor longer on and off the stalk. Normal sweet corn has sugar levels from five to ten percent, SE contains fifteen to 18 percent, and SH2 supersweet has twenty-five to thirty percent. Through the power of hybridization, the new sweet corns provide us with higher yields, and are disease and pest resistant, dependable, and much sweeter for far longer. One of the new sugar-enhanced sweet corns is a 1986 All-America Selections Winner, "How Sweet It Is." It has a delicious flavor, tender eating quality, and retains the sweet flavor even after harvest. Today, there are about 650 open-pollinated varieties of sweet corn, and hybrid varieties number in the thousands.

Over six hundred billion kernels of corn were sold in the United States in 1985. Thanks to the stalwart hybrid varieties, American production has doubled during the last fifty years, and the quest continues for new varieties that will offer greater resistance to insects, disease, and drought and will produce more grain with a higher nutritional value. The future for corn is very bright.

At least it would seem that way. Certain factions in the corn-growing world, however, believe that as we depend more and more on hybrid corn, the open-pollinated varieties will die out. The result, they say, is that nature's stores will become depleted and we will be fully dependent on the man-produced hybrid seeds, which, unlike open-pollinated varieties, do not reproduce true to type from seed and from the parent plant.

## GROWING CORN IN THE HOME GARDEN

If not for their corn crops, the early settlers in Massachusetts and Virginia could not have founded their colonies. Compared to their own slow-growing wheat, barley, and rye, corn was easy to grow, taking only a few weeks to mature.

Corn-growing found favor even among the gentlemen farmers. Benjamin Franklin grew it on his three-hundred-acre farm in New Jersey. George Washington toiled over it in Mount Vernon, New York, and, not to be outdone, Thomas Jefferson had "rows of sweet or shriveled corn" in Monticello, New York. Corn became the subject of much correspondence between Washington and Jefferson.

Although corn is indigenous to the states with a longer, warmer, growing season, it is grown throughout the Northeast and anywhere else that farmers need it for cattle feed, or people want garden-fresh sweet corn. Much of the sweet corn sold in the supermarkets in the Northeast around mid-June comes from Florida. Later on, there is an influx from New Jersey. Unfortunately, by the time these ears reach the dining table, the sweet corn milk has turned to starch. That's why most corn-on-the-cob fanatics make a mad early-morning dash to the local farmer's stand during the season.

Serious corn huskers grow their own, and even small gardens can accommodate small patches. Corn needs at least six hours of sun a day, rich crumbly soil, water, and nitrogen (a high-nitrogen fish emulsion will do the trick).

The Indian women did all the planting in the Northeast, so as to pass along their fertility to the crops they tended. Mounding the earth two or three feet apart, they buried at least one whole fish with the four customary corn seeds in each hill. Squanto, the Indian responsible for showing the Plymouth Pilgrims how to plant corn, was said to chant as he planted the four seeds, "One for the squirrel, one for the crow, one for the cutworm, one for to grow."

The Indian method was to plant rows of beans and squash (*askutasquash*—the Narraganset word for "eaten raw") between the rows of corn. The beans climbed the corn-

# A SAMPLING OF SWEET CORN VARIETIES

*All of these varieties are good for eating on the cob, for freezing, and for canning.*

| VARIETY | DESCRIPTION |
|---|---|
| **SUPER SWEET (SH2) CORN** | |
| Butterfruit | Yellow and also bicolored, this variety is crunchy and delicious to eat raw in salads. |
| How Sweet It Is | White with crisp kernels that stay sweet several days after harvesting. Good for freezing. |
| Early Extra Sweet | Yellow and sweeter than most other corn even after picking. This tender corn is good for freezing. |
| **SUGAR ENHANCED (EH) CORN** | |
| Kandy Korn | A yellow, sweet, high-yielding corn. Although not as crisp as other EH corns, it is still good for freezing. |
| Tendertreat | Yellow with creamy rather than crunchy kernels. Not as starchy as Kandy Korn. |
| Early Glow | Yellow, extra sweet, and extra tender corn. |
| Snow Queen | A very sweet and tender white corn. Holds these qualities after harvesting. |
| Peaches and Cream | Sweet and juicy bicolored kernels. Good for eating raw. |
| **SWEET CORN** | |
| Early Sunglow | One of the sweetest and tenderest of the yellows. |
| Early Golden Bantam | Yellow with large kernels that are sweet and tasty. Particularly good for roasting and freezing. |
| Seneca Chief | Long ears with plump, yellow kernels. Sweet and tender. |
| Golden Midget | Yellow four-inch ears with a high-sugar content. Great for growing in containers. |
| Silver Queen | The queen of corns. A white favorite with large ears and tender, crisp, sweet kernels. Good for freezing. |
| Stardust | A tender, flavorful white that is truly sweet. |
| Platinum Lady | Flavorful, tender, and sweet kernels on long ears. |
| Butter and Sugar | Very sweet and tasty bicolored corn. |

stalks and supplied the soil with nitrogen; the squash shaded the ground, keeping the moisture in and the weeds out. The prickly squash plants were also supposed to discourage raccoons and woodchucks. (Our experience of growing corn and squash together is that the woodchucks love to eat squash leaves and the raccoons still manage to get the corn the night before you plan to pick it.)

What corn you plant will depend on whether you want it for eating fresh on the cob, for popcorn, or for grinding into meal or flour. You might decide that you'd like to grow your own decorative, colored corn, which can also be ground for baking. Black Aztec is an interesting variety to try. It was developed from a strain cultivated in Mexico over two thousand years ago (and is often called Black Mexican). Ready for harvesting two to three months after planting, the corn is at first white. During August, the kernels turn progressively darker until by September they are deep purple. When dried (leave them on the stalks or bring them inside, open up the husks, and hang in a warm, dry place), they turn jet black. At this point, the ears can be left whole and used as ornamentals, or the kernels can be ground into a lavender cornmeal for corn breads or any other dish you would ordinarily prepare with yellow or white cornmeal.

If it's sweet corn you're after, you can choose an all-white variety, like Silver Queen, Chalice, Platinum Lady, or Snow Queen, or a bi-colored corn like Butter and Sugar, Sweet Sue, or Early Gold and Silver. You may prefer a juicy, sweet, and tender yellow corn like Early Sunglow, Tendertreat, or Kandy Korn. Golden Bantam, Truckers' Favorite, and Country Gentlemen are still old favorites with some folks. Sweet corn is not only for eating fresh, it also can be dried and ground into flour.

Corn breeders recommend isolating different varieties of corn by at least 250 feet; separate hybrid sweet whites and yellows and plant yellow (the dominant gene) downwind from white. Isolate extra-sweet hybrid corn from field corn and other types (including regular varieties of sweet corn) by 700 feet. When pollinated by another type, sweet corn develops an abundance of starch kernels. On a small plot of land this problem can be avoided by planting strains of corn that tassel at different times so that they won't be releasing pollen at the same time.

The tassel is the flowering male part of the plant and when "tasseling" takes place, pollen is released into the air. Once airborne, the grains fly in all directions and

You won't be the only one eagerly awaiting corn on the cob. Blackbirds and squirrels sometimes pull up the seedlings to get at the kernels; woodchucks and deer love to eat the tender green shoots. If you have a small patch of corn, deal with these pests by covering it with netting or cheesecloth. Every spring we discourage woodchucks by locating the main entrance to their hole. My daughter and I stand waiting with aluminum pie plates and wooden spoons raised, while my husband puts a hose down the burrow. After about two minutes of running water, out pops the woodchuck to be met by clanging and banging. She runs far away (at least to the neighboring gardens) as fast as her short legs will carry her. Occasionally, a woodchuck will return for a second trauma, but never a third time in one season.

Raccoons are another problem altogether. Like us, they want only the sweetest and ripest ears. Scarecrows and radios do nothing. Some gardeners drop mothballs around the plants to create an offensive odor barrier (don't use those containing paradichlorobenzene). You might also consider trapping the offenders in a Havahart trap and then transporting them to a distant county park. However, your only true defense is to erect an electric fence or train Fido to sit in the corn patch.

There are still other enemies that go by the names of insects and diseases. Here are a few of the most common ones.

**Corn borer:** Borers overwinter in cornstalks; the little blighters start off in the stalks and move into the ripening ears, where they munch on one inch or so of kernels around the tip. As a precautionary measure, shred the stalks in the fall and dig under or compost. Watch for clues: egg masses on the backs of the leaves and tired-looking tassels. To control, spray the tassels with the non-toxic caterpillar-killer *Bacillus thuringiensis* (BT) and import a batch of ladybugs (ladybugs love to feast on these juicy critters). If you notice signs of boring in the stalk, slit it open and remove the grub. The corn is still edible.

**Corn smut:** Corn smut is a ball-like fungus on the ears and stalks. If you spot it, pull up the plant, or affected ear, before the ball bursts and releases thousands of black spores. Burn the offender. In Mexico, this black fungus is eaten as a delicacy. Called *cuitlacoche*, the mushroom-like balls are cut off the cob, sliced and sautéed in garlic and oil until they turn into a soft, black mush. This mixture can be stuffed into tortillas and fried (quesadillas), added to omelettes and crepes, or stirred into sauces and soups. It is in season during the months of August and September.

**Corn rootworm, cutworm, wireworm, white grub, and flea beetles:** All of these insects attack the roots of the plant. One of the best precautions is to buy seed treated with insecticide. If you don't want to do that, try treating the soil with rotonone powder or parasitic nematodes. Buy trichogramma wasps, which lay their eggs in the eggs of cutworms and earworms. Some gardeners sprinkle cornmeal around the seedlings—apparently, the cutworms eat the meal and die from indigestion.

**Corn earworm:** These pests start off on the silks and then burrow into the ear. They lay their eggs on the silks, so at the first sign, spray with BT. Or, as soon as the silks start to appear, brush them with mineral oil once a week.

**Root rot:** Caused by wet cold soil or a fungus in the soil, root rot is one of the two main diseases that the home gardener might anticipate. If it hits, try planting seeds treated with a fungicide the next time.

land on the female silks. There is one silk for every kernel of corn and when a grain of pollen reaches a silk, it travels down the strand to fertilize the ovary. If fertilization takes place, the ovary swells into a plump, juicy kernel.

### PLANTING

Plant corn seeds about a half-inch deep, four to six inches apart and thin eight to ten inches (or follow package directions). To ensure good wind pollination, there should be at least four rows (a total of at least twelve plants) approximately 2½ feet apart, running north to south. Plant as much corn as you can. While most stalks bear one to two ears, some varieties bear three. (Sugar Loaf, a late hybrid corn, is said to bear three, four, and sometimes five ears

per stalk.) Dwarf (midget) sweet corn plants grow about three feet high and bear from three to five ears measuring four to five inches.

To enjoy corn all season, plant several varieties with different maturity dates, or plant the same variety two weeks apart, taking into consideration that some varieties take about 105 days to mature. And don't rush the season. Corn needs temperatures around 55°F. for complete germination.

Besides a sunny location, the plants need mulching with several inches of leaves, straw, or grass clippings. This will control the weeds and keep the large root system moist. If you hill up the rows—draw the soil from between each row towards the plants when they're about three-feet high—the roots will grow stronger.

Mulch will keep the roots moist, but in really dry weather watering will aid tasseling and kernel formation. Soak the soil to keep the roots growing down where they belong.

Rich soil grows the best corn—with bigger ears and tastier kernels. In the spring, dig composted manure, or 5-10-10 or 10-10-10 fertilizer, into the soil to a depth of twelve inches. During the growing season, you'll get good results if you fertilize a couple of times with liquid fish emulsion, 5-10-10, or 12-12-12 (a half-pound per one hundred square feet) after thinning the seedlings and again when the tassels first appear. Rotate corn each year to avoid soil-transmitted diseases.

## HARVESTING

You'll know when sweet corn is ready for picking when the silks start drying out and look brown at the ends—about twenty days after pollination has taken place. Also, the husks should be dark green (your grandmother probably called fresh corn "green corn"). Peel down the husk and look for plump kernels at the tip of the ear. Jab a fingernail into a kernel. If you're right on time, there will be lots of white "milk." If you're too late, there will be very little milk, and if you're too early, milk will be clear. (SH2 and SE high-sugar types will have clear juice.) Remove the ears with a downward pull and twist of the wrist.

Shuck the ears and drop them into boiling water as soon as you can. Indian women kept pots of water boiling alongside the corn field. The faster you get the corn into the pot, the less time the sugar has to turn to starch. There are many schools of thought on boiling corn, but I have only one. Fill a pot with water, bring it to a rolling boil, and drop in the ears for 1½ to 2 minutes. Remove, drain, and roll in butter. Do not add salt to the water when boiling corn, on or off the cob. Salt causes the kernels to toughen.

Some people get a little bit piggy at this point and roll their steaming corn on the cob right into the butter dish. Others feel it's more genteel to place a knob of butter on your plate. Fact of the matter is, no matter how hard you try, you can't look genteel eating something that is held in your hands, drips butter down your chin, and gets stuck in your teeth.

If you don't plan on eating the corn right away, refrigerate it in a perforated plastic bag as soon as possible. Don't remove the husks until you're ready to use it. Once the

*Heap high the farmer's wintry hoard!*
*Heap high the golden corn!*
*No richer gift has Autumn poured*
*From out her lavish horn!*

—JOHN GREENLEAF WHITTIER
THE CORN-SONG

corn is husked, the stored sugar quickly converts to starch.

## DRYING

When growing corn for meal, flour, or popcorn, leave the ears on the stalks until the husks turn brown and the kernels are hard and dry. Remove the ears, pull back the husks, and tie two or three together. Hang from a string strung across a room or store in a basket in a warm, dry place. The kernels should rub off easily when thoroughly dry.

## STORING

Store dried kernels in a glass jar with a tight lid. Keep popcorn kernels in the refrigerator.

If you plan to make your own cornmeal, grind the kernels as you need them. Cornmeal with the germ left in goes rancid quickly, so store it in plastic containers in the refrigerator or freezer for a longer period. How much you plan to grind will help determine the type of mill you'll need. There are several types of grinders available for the job. For small amounts, a hand grinder or blender will do a good job, but if you're grinding for a multitude, you'll be better off with a power grinder. (For further information on manufacturers and sources, see page 25 under "Ground Corn.")

## PUTTING UP FRESH CORN

## FREEZING

When freezing or canning, you need to use the freshest sweet corn you can get your hands on. Although you can remove the kernels with a small, sharp paring knife, working from the top down to the bottom of the ears, there are gadgets on the market that make the removal really effortless, especially if you plan on putting up quarts of kernels or creamed corn.

One corn cutter (available from specialty stores or from seed catalogs) removes the whole kernels cleanly (you don't end up with mashed kernels or bits of cob) and converts to a "creamer" to cut, shred, and scrape the kernels off, leaving the tough skins on the cob.

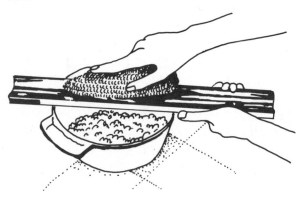

**Whole Ears** Remove the husks and silks from the ears. Heat a large pot of water until bubbles begin to form around the edges, and scald (do not boil) the ears for 5 to 6 minutes. Chill immediately in cold running water, drain, and freeze in plastic bags. I like to use zip-closure bags so I can squeeze all the air out, and I layer them in a single row. That way, the ears don't stick together and I can remove one at a time.

To cook frozen ears: Drop the frozen ears into rolling boiling water and cook 2 to 3 minutes, until heated through or cooked to your liking. Defrosted ears will take 1 to 2 minutes to heat through.

**Whole Kernels** Remove the husks and silks from the ears. Heat a large pot of water until bubbles begin to form around the edges, and scald (do not boil) the ears for 1–2 minutes. Chill in cold running water and drain.

Remove the kernels with a small, sharp knife, starting at the top and working down to the bottom. Package in airtight containers or plastic bags and freeze. Cook for 2 minutes if defrosted and 3–5 minutes if frozen.

**Creamed Corn** Remove the husks and silks from the ears. Using a small, sharp knife, cut down the center of each row of kernels, and, working from top to bottom, scrape the kernels into a bowl.

Measure the kernels by the cup and place in a large pan. For each cup of kernels, add 1 teaspoon cornstarch and combine. For each cup of kernels, pour ¼ cup low-fat milk into a pan and bring to a boil. Stir the boiling milk into the kernel and cornstarch mixture. Stirring frequently, simmer the creamed corn for 5 minutes, until it thickens.

Spoon into plastic freezer containers, leaving a half-inch of head room, and freeze. Cook frozen or thawed creamed corn over gentle heat (or in the microwave) until heated through.

**Unblanched Corn in Husks** This method is taken from *The Busy Person's Guide to Preserving Food* by Janet Bachand Chadwick, another title from Garden Way Publishing.

If you don't have your own garden crop, get the freshest corn possible from your favorite orchard or farmers' market.

Leave the husks on the ears of corn and lay them in a single layer on a cookie sheet. Freeze for 48 hours. Place the frozen ears in plastic freezer bags and freeze for up to four months.

"To cook corn frozen in husks," Mrs. Chadwick recommends, "husk the frozen corn under cold running water. Remove the silks by rotating your hands around the corn. Put the corn in a pan, cover with cold water. Cover the pan, heat to boiling, and boil for 1 minute. Remove from heat, and let the corn stand in hot water 5 minutes. Eat immediately."

## CANNED WHOLE KERNEL CORN

Using a sharp knife and working from top to bottom, cut the whole kernels from the cobs. Place the kernels in sterilized pint jars and leave one inch of head space. Cover with boiling water, again leaving one inch of head space. Put lids on the jars.

A pressure cooker must be used to can corn, so, following the manufacturer's directions, prepare the canner and process the pint jars at 10 pounds pressure for 55 minutes. Test lids for a proper seal.

## CORNMEAL

Thousands of years ago, the only type of corn grown by Indians living in the Americas produced kernels encased in a tough skin. To render the corn edible, the kernels had to be ground into meal. Thus, ground corn became a staple food in those early cultures.

Originally the corn was ground between stones, and to make this an easier task, the Indians first treated the corn to soften the hulls. In one method, the corn was briefly boiled, dried, and then parched by roasting the kernels. Other times, the hard hulls would be dissolved by soaking them in wood ashes and water. In the Southwest, Indians ground the corn between soft sandstone, and the grit that was ground off the stone into the corn caused their teeth to wear down prematurely. The settlers also used stone, but in the form of huge grinding wheels that were powered by hand, water, and wind. Steel rollers replaced the stone wheels in the mills around the 1850s.

Some mills still produce stoneground corn-meal. It is not only tastier, but is also more nutritious than that put through the steel rollers. Stoneground corn contains the whole kernel, whereas steel rollers eliminate the corn germ. (The germ, which contains the oil, is made into corn oil.)

Ground corn is still a staple food in countries in South and Central America and South Africa. To many people, Mexican food is

---

## GRAIN GRINDERS

There are several hand and electric grinders on the market. The hand grinders do a good job; they just take time and muscle power, which is fine if you grind only small batches at a time. With corn, you don't want to grind too much at once because cornmeal goes rancid quickly.

Electric power mills are available from the following sources and can cost up to $200.

*Excalibur Flour Mill*
Excalibur Products
57711 Florin-Perkins Road
Sacramento, CA 95828

*The Kitchen Mill*
Williams Sonoma Inc.
Mail Order Department
P.O. Box 7456
San Francisco, CA 94120

There is a new electric grinding machine called *The Kitchen Mill* available, and it has been highly recommended by several people. Texture can be controlled by a dial, and the grains are ground quickly to a smooth flour. If you're grinding wheat or rye, for example, this is most desirable.

*Marathon Flour Mill*
The Grover Company
2111 South Industrial Park Avenue
Tempe, AZ 85292

*Magic Mill*
235 West 200 South
Salt Lake City, UT 84101

*Mill-Rite Flour Mill*
Retsel Corporation
Box 47
McCammon, ID 83250

synonymous with cornmeal tortillas and tamales. However, corn is integrated into the
Mexican diet in numerous forms and is the
basis of many dishes. In the Southwest, the
Hopi Indians, who inhabit ancestral pueb-
los on top of mesas in Arizona, still rely on
corn for their main supply of starch. As
their ancestors did, they dry surplus ears,
stacking and storing them according to the
different colors and varieties. Some of the
dried kernels are soaked, then simmered
with beans (for succotash), squash, peppers,
and/or meat. Others are ground into a meal
by the ancient method between stones, then
mixed with water to form a batter for making *piki*, a thin bread cooked over a hot,
greased stone.

## GRITS

In the Southern United States, hominy,
the name for bleached and skinned whole
corn kernels (called *posole* in the southwestern states) is eaten daily in the form of
grits, a coarsely ground cornmeal. Grits are
served at breakfast along with ham and
eggs.

Grits go back to 1607, when members of
the London Company came ashore in Jamestown, Virginia. They were welcomed by
the local Indian tribe with bowls of "rock-
ahominie"—a mush of hot maize. It tasted
good to the hungry seafarers and was
adopted by their community. They learned
to grind it a little finer and called it "hominy grits." Since that time grits has become
a mainstay of southern cuisine—more than
two-thirds of the 150-million pounds of grits
milled annually in the United States is consumed in southern states. Grits on its own
is simply mealy mush, but add a little butter and salt and pepper and it already takes
on character. In the South, there's grits-
and-gravy, grits-and-eggs, grits-with-meat,
grits-and-cheese, grits-and-sausage, grits-
and-grillades (sautéed or grilled meat with
herbs), and many more combinations.

Grits were first served at the President's
table to Ulysses S. Grant, who had developed a liking for the dish during his so-
journ in the South from 1861 to 1865. When
Jimmy Carter was President it is said that
he ate grits for breakfast every day.

Grits are available at most large supermarkets and specialty grocery stores throughout the United States.

## TYPES OF CORN

The cornmeal you choose makes a huge
difference to what you are making. Take
johnny cakes, for example. Rhode Islanders
eat johnny cakes at just about every meal.
They may be served like pancakes for break-

fast, or with meats and other main courses in place of potatoes, and then there are dessert johnny cakes. It's been this way for the last three hundred years or so. Unfortunately, Rhode Islanders are split into two factions as to what is the authentic recipe— the Newport County method with cornmeal, salt, and milk mixed to a runny consistency, or the South County method of mixing cornmeal with boiling water and salt to a thicker consistency. Then there are those staunch supporters of the Society for the Propagation of the Johnnycake Tradition in Rhode Island, who believe that only the use of flint corn (a now obsolete strain of corn that originated from maize grown by the Narraganset and Wampanoag Indians) is traditional and entitles the cakes to be called Rhode Island johnny cakes.

White and yellow are the common colors associated with cornmeal, and Yellow Dent, which is grown in the central and southern corn belts, and Longfellow Flint, grown in the northern states. Hickory King is a favorite white for hominy, grits, and the finer *masa harina* corn flour used for tortillas and thickening stews. Some multicolored corn is also used for cornmeal, flour, and hominy in the Southwest.

All cornmeal is nutritious (along with oatmeal, it is second to wheat in protein content), and it is a boon to people who are allergic to the protein in wheat gluten. When cornmeal is used without the addition of wheat or gluten flour, the results are dense and coarse. However, a lighter product can be obtained by substituting ⅓ to ½ of the cornmeal with all-purpose flour or, to a lesser degree, with rye or whole wheat flour. Baking powder or baking soda will help baked goods to rise. If still more volume is desired, well-beaten egg whites will increase the air content.

## CORNSTARCH

Cornstarch is obtained from the endosperm layer of the kernel and is called corn flour in Europe, where it is used to a greater extent in home cooking than in American kitchens. However, commercial establishments in America use it in all manner of baked goods, canned gravies, and baking powder. I use cornstarch in place of flour if I'm thickening stews, fruits, or sauces, and in some puddings (see Chocolate Pudding on page 151). Mix 1 tablespoon cornstarch with 1 tablespoon water to a smooth consistency and then add to the hot mixture. One tablespoon will thicken 1 cup of liquid. If a recipe calls for 4 tablespoons (¼ cup) of flour, substitute 2 tablespoons of cornstarch. When thickening soups or stews, add the cornstarch (in a drop of water) during the last 5 minutes of cooking. Overcooking cornstarch causes it to thin out.

## CORN SYRUP

Corn syrup is made from cornstarch (as is laundry starch) and is used in the confectionery trade to give a smoother texture to frostings, chocolate, and desserts. The light syrup, in particular, is used to produce clear jams and jellies, and the syrup for canned fruits. In fact, there are not many food products on the supermarket shelves that do not contain corn syrup. It's in breads, cakes, cookies, cereals, and many other products.

## BLUE CORN

The Navajo and Hopi Indians of the southwestern states have grown blue, black, and red corn for centuries. The blue corn, though, has been the favorite for grinding into cornmeal, for their daily needs. The Indians say that blue corn gives strength, and those that are sick often eat a blue cornmeal mush (*atole*) or soup two or three times a day. Others dilute it with water and drink it. It is also made into *piki* breads.

During the last five years blue cornmeal has become a popular commercial commodity and has been slowly invading health food and gourmet stores across the nation. There were only five hundred acres of blue corn grown in 1983; in 1986 there were close to fifteen hundred acres—still a miniscule amount compared to the 620,000 acres of sweet corn that is grown.

The kernels of blue corn are deep blue with a purple cast and turn gray-blue when ground. When mixed with liquid, the color is lavender rather than blue. Due to the high starch content, blue cornmeal is even more dense than yellow cornmeal, and when made into a pancake mix, it is usually combined in equal proportions with wheat flour to produce a lighter texture. Otherwise, blue cornmeal can be substituted for yellow and white cornmeal in most recipes.

Tortilla factories are the major processors of blue corn, but there are smaller producers who supply stores with cornmeal, pancake mixes, and whole dried kernels. These include Arrowhead Mills of Hereford, Texas, and Ross Edwards' Blue Corn Connection, 8812 4th Street, N.W., Albuquerque, New Mexico 87114, (505) 897-2412. Ross Edwards also has a mail order business for his Blue Heaven Pancake Mix, and his tortilla chips. In New York, Jane Butel of Pecos River Spice Company packages and distributes blue cornmeal to specialty stores and gourmet food sections in the larger department stores.

# POPCORN

You might think of popcorn as something to eat in the dark at the movies or while sitting in front of a TV set. In actual fact, eating popcorn has been in vogue for several thousand years—not smothered in butter or caramel but coated in the ashes from an Indian's fire.

## THE FIRST POPCORN

The precursor of sweet, flint, and dent corn, popcorn was cultivated by the Incas, Mayas, Aztecs, and North American natives. These Indian cultures used popcorn for food, barter currency, necklaces, and decoration. Magical powers were attributed to popcorn. It was strewn about doorways as a sign of hospitality and to ward off enemies. It was also a symbol of fertility.

The Indian who first dropped a kernel or cob at the edge of the fire and witnessed the hard seed exploding inside out to fluffy white softness must have thought it magic, indeed. To early Indians without grinding tools, this method turned the hard-hulled seeds into an important food source.

Popcorn is the only type of corn that truly explodes when heated. Its hard coat contains a core of soft starch that expands when heat is applied. Popcorn comes in white, yellow, black, multicolored, and strawberry red. (These are the little red ears that look something like large strawberries and make pretty decorations on the Thanksgiving table.) Dried sweet corn will also puff somewhat when heated and is usually referred to as parched corn or puffed corn.

As the story goes, the early settlers caught on to popcorn when Quadequina, an Algonquin Indian, contributed a deerskin bag of popped strawberry corn to the Thanksgiving feast. It turned out to be more than a welcome gift. The Pilgrims learned how to "parch" their own corn and served it with milk as a breakfast cereal. They also borrowed the Indian method of making *nookick*. Corn was popped in earthenware pots placed in the hot ashes, pulverized, and taken along on journeys. During the trip it would be reconstituted with water to make a nutritious meal. In colonial days, a metal perforated cylinder was used to pop the corn. Set next to the fire, it was turned by an axle. Fifteen hundred years ago, Indians in both South and North America used large (up to 8 feet wide), shallow clay or metal pots (sometimes with tripod legs) over a fire. This method of popping is still used by the Papagos in Arizona. The Winnebago Indians of Illinois have always popped the whole corn on the cob by spearing it with a sharp stick and holding it near the fire.

## POPCORN TODAY

Today's popcorn may be much more fluffy and tender than that eaten in colonial times, but it's still basically the same complex carbohydrate food that Indians were eating thousands of years ago. This is why it makes such a great healthy snack—1 cup of plain popcorn contains only twenty-five calories (fifty with butter or oil). Four cups of this high-fiber food can leave you feeling quite satisfied.

Popcorn left the hearthside towards the end of the nineteenth century to be peddled on the streets and at carnivals. But it was Charles Cretors who turned it into the fad it is today. In 1891 he took his popcorn business to the streets in a handsome shiny red cart. He also developed the method of popping with oil. During this time someone else was selling a sticky mixture of popcorn and peanuts coated with a chewy molasses. In 1896, F. W. Rueckheim of Chicago gave his concoction the brand name Cracker Jack, and it's still going strong today.

Popcorn's future was secured with the advent of the movie house. Today that business accounts for over $300 million in one year. Americans snack on over six million pounds of popcorn per year (about forty quarts per person), seventy percent of which is popped at home. There are stores that sell nothing but popcorn, and some sell more than thirty flavors, including chili, pizza, strawberry, and chocolate.

## POPPING CORN AT HOME

When you're popping at home, the inevitable question arises: "What's the best popper?" The answer is: "Whatever you like." There are those who prefer the clean method of electrically operated hot-air machines that don't use oil (great for dieters) and have a fan that blows the popped corn into a bowl. There are also electric machines that work with oil. For those of you who don't like new-fangled machines, you can still buy stove-top, hand-cranked poppers, or you can get excellent results from a heavy skillet with a good-fitting lid. I used this method for years but switched to a hot-air machine so that my daughter Wendy could make her own popcorn when she was six years old. Now that she's nine, she also uses the microwave oven.

When popcorn kernels don't pop, they're

probably too dry. These duds can be saved. Sprinkle ¼ cup of kernels with 1 tablespoon water, cover, and let sit for at least an hour. Drain off any excess water and pop them without oil (water hitting hot oil causes enormous spattering). The best place to store the kernels is in a glass jar in the refrigerator—this prevents moisture loss.

Now, if you're already on your way into the kitchen to do some popping, you might like to try some of the recipes on pages 156–158.

## CORN NUTRITION

The Indians called corn *maize*, meaning "our life," because it sustained them from harvest to harvest and without it they would not have survived. (It came by its botanical Latin name *Zea mays*, "that which sustains the Mayas," by way of the Swedish botanist Linnaeus, in 1737.)

Although in Latin America and South America, corn is a staple food crop, it is considered an incomplete protein and needs supplementing to create a complete protein meal. This can be done by combining corn with legumes (beans, peas), nuts (including peanut butter) and seeds, or small amounts of animal protein such as milk, cheese, yogurt, eggs, meat, fish, or poultry. In many of their dishes the Mexicans combine corn with rice or beans or a dairy food, such as sour cream or cheese. When we eat corn muffins for breakfast, we are eating a complete protein because we have added milk and eggs to the batter.

Corn contains phosphorous, iron, calcium, niacin, and riboflavin and provides roughage from the cellulose hulls. It is also a source of beta-carotene (a natural form of vitamin A) and vitamin C; yellow corn is a better source than white corn. A corn kernel is composed of approximately 71 percent carbohydrates, 10–14 percent water, 10.5 percent protein, 3 percent fat, and 1.5 percent fiber.

One medium ear of fresh sweet corn is about seventy calories (hold the butter, please), the same as a half cup of plain kernels.

So, in the knowledge that every morsel (or kernel) you consume is contributing to your health and well being, enjoy yourself as you sample and experiment with the delicious recipes that follow.

# EASY
# SALADS

CHAPTER 2

*First the blade, then the ear, after that the full corn in the ear.*

—MARK 4: 28

I serve my family a salad every evening, usually in addition to a vegetable dish. When there's a surplus of fresh vegetables in the summer, I quick-cook the beans and bell peppers, pour a vinaigrette-type dressing over them, and chill for future use. This mixture becomes a base for a different type of salad. Sometimes I add sliced beets and corn, grated carrots and corn, peas and corn, or anything else I think will look and taste good together. In the winter, when there's a dearth of fresh vegetables, I cook up tons of beets before the ground freezes and resort to tiny frozen peas and canned corn kernels. (There are some pretty good canned corn kernels—just have a tasting test until you find the crunchiest. The same goes for frozen corn kernels, although I've found the canned to be superior—just the opposite of peas.)

If you're a salad lover like me, your family probably gets a salad packed into their lunch boxes. Or, if you work at home, you sit down to several salad lunches a week. This is when it's handy to have canned corn on hand at all times. Open up a can and mix with leftover rice, barley, tortellini, small pasta shapes, lentils, kidney beans, or even potato salad. Instead of making a tuna sandwich, combine the drained tuna (I use tuna packed in water) with ½–¾ cup corn kernels, 2–4 tablespoons sour cream, and ½ teaspoon dried dill or tarragon. Lightly toast one or two slices of bread per person and spoon the salad on top.

One of my favorite lunches is to combine 1 cup corn kernels with ¼ cup cottage cheese and some snipped chives. If you add ¼ cup kidney beans (or any other dried beans or nuts), you'll be getting a complex carbohydrate, complete protein meal.

# CORN WITH ZUCCHINI AND RED BELL PEPPER

PREPARATION: 10–15 MINUTES      COOKING: 6 MINUTES      YIELD: 4 SERVINGS

2 tablespoons butter blend or vegetable
   oil
1 medium red bell pepper
2 small zucchinis, diced small
1 clove garlic, crushed or minced
½ teaspoon coarsely ground black pepper
1½ cups corn kernels (3 ears of corn)
½ cup sour cream

**1.** Heat the butter in a large skillet and sauté the red pepper for 1 minute.

**2.** Add the zucchini, garlic, and black pepper and sauté for 2 minutes.

**3.** Stir in the corn, cover the skillet, and cook for 2 more minutes.

**4.** Spoon the mixture into a warm serving dish, combine with the sour cream, and serve immediately or chill.

# CORN WITH NUTS

PREPARATION: 10 MINUTES      YIELD: 2 SERVINGS

½ cup low-fat, small curd (or creamed)
   cottage cheese
2 teaspoons sesame oil
2 teaspoons soy sauce
¼ teaspoon hot oil (optional)
2 cups corn kernels
¼ cup raisins
¼ cup sesame seeds (toasted, if desired),
   sunflower meats, or chopped peanuts

**1.** In a medium mixing bowl, combine the cottage cheese, sesame oil, soy sauce, and hot oil.

**2.** Stir in the corn kernels.

**3.** Place each serving on a plate and sprinkle raisins and sesame seeds over the top.

# CORN WITH BEANS

PREPARATION: 10 MINUTES                 YIELD: 2 SERVINGS

4 slices bacon
¼ cup sour cream
¼ cup plain yogurt
¼ cup chopped chives or sliced scallion
   greens
2 cups corn kernels, raw or cooked
½ cup cooked kidney or pinto beans
4 lettuce leaves

**I.** Fry the bacon until crisp. Drain and crumble.

**2.** Using a medium mixing bowl, combine the sour cream, yogurt, and chives.

**3.** Stir in the corn and beans.

**4.** Serve on a few lettuce leaves and sprinkle with the crumbled bacon.

# HAM AND CORN SALAD

PREPARATION: 10 MINUTES                 YIELD: 4 SERVINGS

*I also use leftover chicken for this, but any cold meat will do.*

1 cup olive or vegetable oil
2 tablespoons apple cider or wine vinegar
2 tablespoons lemon juice
2 teaspoons dried basil, or 1 teaspoon
   chopped fresh ginger
1 clove garlic, crushed
1 teaspoon prepared mustard
1 red onion, sliced very thin
1 stalk celery, sliced very thin
2 cups ham, diced in ½-by-1-inch pieces,
   or cooked chicken, diced in 1-inch pieces
1 cup corn kernels
2 Bibb lettuces, washed and dried

**I.** Place the oil, vinegar, lemon juice, basil, garlic, and mustard in a jar and shake vigorously to blend.

**2.** Combine the onion, celery, ham, and corn in a large bowl and toss with the salad dressing.

**3.** Arrange lettuce leaves on a large serving platter and spoon the ham mixture on top.

# RICE, LENTIL, AND CORN SALAD

PREPARATION/COOKING: 30 MINUTES          YIELD: 8 SERVINGS

*As someone who is more comfortable eating vegetables, legumes, and grains than meat, this is one of my favorite meatless, complete protein dishes. Because I cook lentils and rice frequently, I usually have them precooked in the freezer or refrigerator so my preparation time is shorter than the one given below. If you don't have the time to cook brown rice, use the parboiled varieties; for the lentils you can substitute canned black beans. This recipe makes an interesting combination for the palate and the eye.*

1 cup dried lentils (3 cups cooked)
1 cup brown rice (3 cups cooked)
¼ cup apple cider vinegar
¾ cup olive oil
1–2 large cloves garlic, crushed
1 tablespoon dried basil
2 teaspoons prepared mild mustard
1 large red bell pepper, diced small
4 large scallions, including the greens,
   sliced thin
3 cups corn kernels

**1.** Rinse the lentils, drain, and place in a 2-quart saucepan with 4 cups of water. Bring to a boil, cover the pot, and simmer for 20–30 minutes, until just tender. Remove and drain.

**2.** Sprinkle the rice into 2 cups of boiling water, cover the pot, and simmer for 20–30 minutes, until tender and the water has been absorbed.

**3.** While the rice and lentils are cooking, prepare the salad dressing. Place the cider vinegar, olive oil, crushed garlic, basil, and mustard in a 2-cup jar and shake vigorously.

**4.** Rinse the cooked rice under cold water and combine with the cooked lentils in a large bowl.

**5.** Add the bell pepper, scallions, and corn to the rice and lentils. Toss together with the salad dressing. Chill if desired, or serve at room temperature.

# CORN-STUFFED TOMATOES

PREPARATION: 20 MINUTES          BAKING: 15 MINUTES          YIELD: 6 SERVINGS

*These can be served as a baked hot side dish, or prepare them up to the point of baking and serve cold as part of a luncheon salad.*

6 large, ripe (but firm) tomatoes
4 ears of corn (or 2 cups corn kernels)
⅓ cup chopped chives or scallion greens
1 red bell pepper, chopped fine
½ teaspoon coarsely ground black pepper
2 tablespoons mayonnaise
6 tablespoons grated cheddar cheese

**1.** Preheat the oven to 400°F., if you are serving the dish hot.

**2.** Remove a ¼-inch slice from the stem end of the tomato and scoop out the seeds with a spoon.

**3.** Turn the tomato shells upside down and drain on paper towels.

**4.** Working from top to bottom of the ears of corn, use a sharp knife to cut off the kernels.

**5.** Place the kernels in a medium mixing bowl with the chives, bell pepper, black pepper, and mayonnaise. Combine.

**6.** Stuff into the tomatoes and sprinkle 1 tablespoon grated cheese on top of each one.

**7.** Chill and serve as a salad, or bake in a preheated 400°F. oven for 15 minutes and serve as a hot side dish.

# SOUPS & CHOWDERS

## CHAPTER 3

*A land of corn and wine.*

—2 KINGS 18: 32

These dishes are particularly appealing in summer when the corn is fresh. However, I don't hesitate to make them year-round with canned or frozen corn. For eye-pleasing combinations, serve them sprinkled with chopped fresh herbs and side salads of red peppers and spinach, vine-ripened tomatoes, mozzarella, and fresh basil. Have a basket of fresh-baked muffins on the table, and you'll make your family and guests very happy. Bon appétit!

# QUICK CORN CHOWDER

| PREPARATION: 5 MINUTES | COOKING: 20 MINUTES | YIELD: 4 SERVINGS |
|---|---|---|

*Whenever I have left over boiled or mashed potatoes, I toss them into the soup pot, fix hash, fry up potato cakes, or make them serve as a pie crust. In fact, I usually make extra so that I am sure of having leftovers. The same goes for creamed corn. This can go into soufflés, puddings, omelets, pancakes, and muffins, and, of course, chowder.*

4 slices bacon (optional)
1 tablespoon vegetable oil
1 medium onion, chopped
2 cups mashed potatoes
2 cups creamed corn (or use corn kernels
   and purée ½ cup)
½ teaspoon coarsely ground black pepper
1 teaspoon curry powder
4 cups low-fat milk

**1.** Heat a large kettle and cook the bacon until crisp. Remove and drain on paper towels. Discard the fat.

**2.** Heat the oil and sauté the onion until lightly golden, about 4 minutes.

**3.** Stir in the rest of the ingredients, cover partially, and simmer over low heat for 15 minutes or until hot.

**4.** Sprinkle a little crumbled bacon on top of each serving.

# CORN AND SHRIMP SOUP

PREPARATION: 5–10 MINUTES          COOKING: 10 MINUTES          YIELD: 2–4 SERVINGS

*This soup is absolutely delicious. For a richer broth, substitute 1 cup low-fat milk for 1 cup of the stock. Sometimes I add 1 teaspoon of curry powder.*

2 tablespoons butter blend
1 large onion, sliced thin
3 ears of fresh sweet corn, or 1½ cups kernels
2 cups chicken stock
1 teaspoon dried thyme
¼ teaspoon dried tarragon
½ teaspoon coarsely ground black pepper
1½ pounds medium shrimp, shelled and deveined
¼ cup chopped parsley
½ cup croutons (sauté in 1 tablespoon butter blend, if desired)

**1.** Heat the butter blend in a large skillet and add the sliced onion.

**2.** Sauté for 4–5 minutes over very low heat, until golden.

**3.** Using a sharp knife and working from top to bottom, remove the corn from the cobs.

**4.** Place in a blender or food processor with the cooked onion, stock, thyme, tarragon, and pepper.

**5.** Blend or process until smooth.

**6.** Pour into the skillet and heat for 2 minutes until steaming.

**7.** Add the shrimp and cook over low heat for 2–3 minutes.

**8.** Spoon into bowls and sprinkle with chopped parsley and croutons.

# CORN, BEAN, AND SQUASH SOUP

PREPARATION: 15 MINUTES        COOKING: 35 MINUTES        YIELD: 4–6 SERVINGS

*For a wintertime soup, use butternut squash or pumpkin. Vary the flavors by adding 1 teaspoon each of chili powder and cumin.*

2 tablespoons vegetable oil
1 large onion, chopped
2 cloves garlic, minced or crushed
2 8-inch zucchinis, sliced ½ inch thick (or 2 cups cubed and peeled butternut or pumpkin)
2 cups chicken stock
2 cups chopped tomatoes, fresh or canned
1 teaspoon dried oregano
1 teaspoon dried thyme
1 teaspoon dried basil
2 cups cooked or canned pinto or red kidney beans
2 cups corn kernels, fresh, canned, or frozen

**1.** Heat the oil over medium in a large skillet and sauté the onion for 3 minutes, until golden.

**2.** Add the garlic and zucchini and cook 5 minutes.

**3.** Add the stock, tomatoes, oregano, thyme, basil, and beans. Simmer for 15 minutes.

**4.** Add the corn and cook 10 minutes longer (or, if using winter squashes, until they are tender).

# CORN AND RED PEPPER CHOWDER

PREPARATION: 20–30 MINUTES          COOKING: 40 MINUTES          YIELD: 6 SERVINGS

*When a soup contains potatoes and milk, it is usually called a chowder. Chowder comes from the French* chaudière—*a cauldron in which Breton fishermen used to stir up a mess of fish stew.*

2 tablespoons vegetable oil
2 leeks, washed and sliced, OR 10 scallions, sliced
2 carrots, chopped fine
2 red bell peppers, diced
4 medium red potatoes, peeled and cut in 1-inch dice
3 cups chicken stock
2 cups low-fat or regular milk
1 bay leaf
1 teaspoon dried thyme
½ teaspoon coarsely ground black pepper
8 ears of corn, or 4 cups corn kernels
¼ cup chopped parsley

**1.** Heat the oil in a 4–6-quart Dutch oven. Add the leeks and carrots and sauté over low heat for 5 minutes without browning.

**2.** Stir in the red peppers and sauté 3 minutes longer.

**3.** Add the potatoes, chicken stock, milk, bay leaf, thyme, and black pepper. Cover partially and simmer for 20 minutes, until the potatoes are tender but not falling apart.

**4.** Using a sharp knife and working from top to bottom, remove the corn kernels from the cob. Scrape off all the pulp.

**5.** Crush about one-third of the cooked vegetables and add the corn to the kettle. Simmer for 10 minutes.

**6.** Sprinkle parsley on top of each serving.

# JOHN ATWOOD'S CURRIED CORN CHOWDER

PREPARATION: 20 MINUTES          COOKING: 65 MINUTES          YIELD: 8 SERVINGS

*A friend of the family, John Atwood, is a joy to have to dinner. He eats his food with gusto and is young enough to be able to eat a lot of what he wants. John also likes to cook and when he does, he shares it with friends. Talking about his chowder, John said "The flavors increase as it ages, but it tastes so good you won't be able to resist eating it the first day. So make twice as much. Also, it's hot so I accompany it with beer, and that's got to be Eagle beer!"*

4 medium Idaho baking potatoes, cut in
  ½-inch cubes
2–4 tablespoons butter
8 large onions, sliced thin
2 cloves garlic (optional)
2 tablespoons curry powder
3 cups corn kernels
1 cup light cream

**I.** Put the potatoes in an 8–10-quart Dutch oven and pour in enough water to cover 2 inches above.

**2.** Bring to a boil, then simmer for 15 minutes, or until almost tender.

**3.** Heat the butter in a large (13-inch cast-iron) skillet and add the onions and garlic. Cook gently for 5 minutes.

**4.** Stir in the curry powder and cook 5 minutes.

**5.** Add the onion mixture and the corn to the potatoes. Simmer for 30 minutes.

**6.** Add the cream and simmer for 10 minutes longer.

# PENNSYLVANIA DUTCH CHICKEN-CORN SOUP

1 3½–4-pound chicken, cut up with skin removed
1 large stalk celery, chopped
1 large onion, chopped
10 cups chicken stock, or water with a bouillon cube
1 pound wide egg noodles
4 cups corn kernels
½ teaspoon coarsely ground black pepper
¼ teaspoon saffron
¼ cup chopped parsley
2 large eggs, hard boiled, cooled and chopped

**1.** Place the chicken pieces, celery, onion, and stock in a 6-quart kettle. Bring to a boil rapidly; skim off the scum that rises to the surface.

**2.** Reduce the heat and simmer, partially covered, for 30 minutes, until the chicken is tender.

**3.** Transfer the chicken to a plate and remove the meat from the bones.

**4.** Return the stock to a boil and add the noodles. Cook 10 minutes.

**5.** Return the chicken pieces to the pot with the corn, pepper, saffron, and parsley. Cook 5 minutes.

**6.** Spoon into serving bowls, sprinkle with the chopped egg, and serve.

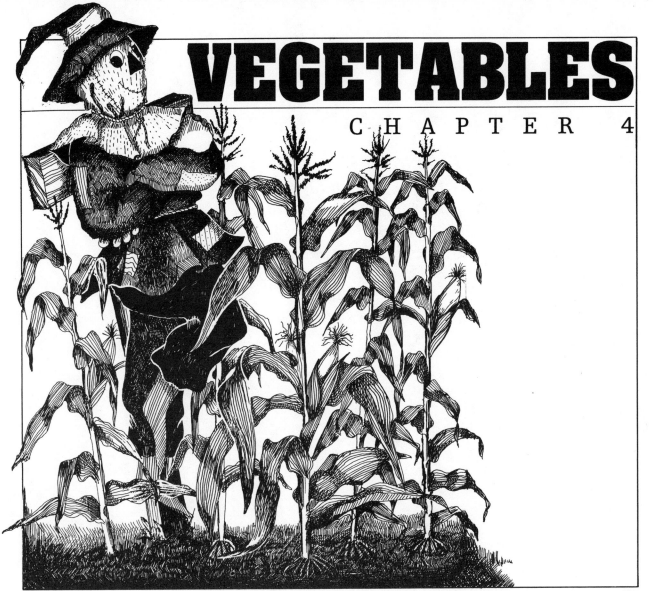

# VEGETABLES

## CHAPTER 4

*She stood breast-high amid the corn,*
*Clasped by the golden light of morn,*
*Like the sweetheart of the sun*
*Who many a glowing kiss had won.*

—THOMAS HOOD (1798–1845)
RUTH

**A**lthough many people consider it at its best simply dropped in boiling water and rolled in butter, corn has a distinctive flavor that combines well with a mixture of ingredients. One of my favorite combinations is a cup of chopped red bell pepper sautéed in olive oil for 3 minutes and 2 cups of corn kernels (fresh or canned) stirred in and cooked for 2 minutes, until the mixture is hot. I vary the flavor of this dish with ½ teaspoon of basil or thyme, or ¼ cup of chopped chives.

Corn can be mixed into just about any vegetable dish. It goes well with just one or two vegetables like tomatoes and squash, peas and beans, or sweet peppers and onions; it adds color, texture, and flavor to a whole mess of vegetables combined, such as onions, zucchini, peppers, tomatoes, and eggplant.

When we don't get around to eating as many ears of corn as I think we will, I keep the extra ears unhusked in the refrigerator until the next day. At that time, I husk them and scrape the corn off the cob. For every 2 cups of corn kernels, I melt 1 tablespoon of unsalted butter (blend or margarine) and heat the corn for about 3 minutes over gentle heat. Then I sprinkle it with pepper, and everyone salts to taste. Just one day old, the corn is sweet with a super-crunchy texture.

# CORN ON THE COB

PREPARATION: 1–2 MINUTES      COOKING: 1½–2 MINUTES PER BATCH      YIELD: 4 SERVINGS

*This is the easiest way of cooking corn, but all too often the crunchy, delicious ears are overcooked. It's all very personal, of course, but when corn is sweet and crisp enough to eat raw, all it needs are a couple of minutes in boiling water to heat it through.*

*Count at least two ears per person and cook the second serving while you're eating the first.*

**4 quarts water**
**8 ears of sweet corn**

**1.** Using a pan large enough to hold 4 ears of corn without crowding, bring the water to a boil.

**2.** While you're waiting for the water to boil, shuck the corn and remove the silks.

**3.** Drop 4 ears of corn at a time into the boiling water, and cook for 1½–2 minutes. Remove with tongs and serve immediately.

**4.** Allow the diners to butter and season their own corn.

**5.** Cook the second batch of corn while eating the first.

## Steamed Corn:

*Place ½ cup of water in a large kettle, add the ears of shucked corn and cover. Bring the water to a boil and cook for 5 minutes. Remove the cover and serve the corn immediately.*

# BARBECUED CORN

*This makes an interesting change from boiled corn and certainly simplifies things if you're having friends over.*

**8 ears of corn**
**8 tablespoons butter (1 stick)**
**½ teaspoon coarsely ground black pepper**
**1 clove garlic, crushed or minced (optional)**
**8 sprigs fresh basil or other fresh herb**

**1.** Husk the corn.

**2.** Mash the butter, pepper, and garlic together in a bowl.

**3.** Spread the corn with the butter mixture on one side and place butter-side down on individual pieces of aluminum foil.

**4.** Spread the top side of the corn with more mixture and add a sprig of basil. Wrap tightly in the foil.

**5.** When the barbecue coals are gray (about 15–20 minutes after lighting), place the foil-wrapped ears on the grill 4 inches from the heat.

**6.** Grill for 15–20 minutes, turning the packets every 4–5 minutes.

We tend to be corn purists in this family in that we like the unadultered corn-on-the-cob flavor. However, sometimes it's in order to roll the corn in flavored butter. To every stick of butter (½ cup or 8 tablespoons) add 1 teaspoon of dried herbs such as oregano, basil, or tarragon. Fresh herbs also can be used. Try 2 tablespoons of chopped chives or parsley; or 1 tablespoon of chopped dill, basil, thyme, oregano, or tarragon. Add salt and pepper as desired.

# ROASTED CORN

PREPARATION: 15–20 MINUTES     COOKING: 15–20 MINUTES     YIELD: 4 SERVINGS

*This is the way the Indians cooked corn, although they would put the ears directly in the hot ashes to roast, or cover them with a layer of earth and then pile the hot coals on top. To shuck the corn while it's hot, hold the ears in an old kitchen towel if necessary, OR, to almost eliminate this procedure, pull back the husks before roasting and secure them with strips of foil.*

**8 ears of corn, unshucked**

**1.** Place the unshucked ears of corn on top of the barbecue grill, 4 inches from the heat.

**2.** Grill for 10 minutes, turning them every 2–3 minutes.

**3.** Remove the husks, return the corn to the grill, and continue to roast for 8 minutes more. Give them a quarter turn every 2 minutes.

**4.** Serve immediately with butter and seasonings.

# CORN AND TOMATOES

PREPARATION: 10 MINUTES          COOKING: 20 MINUTES          YIELD: 6 SERVINGS

*This is a quick way to use up surplus garden tomatoes and leftover corn on the cob. I like to serve this over rice with broiled bluefish, cod, or halibut steaks.*

2 tablespoons olive or vegetable oil
1 large red Bermuda onion, sliced paper thin
2 cloves garlic, minced or crushed
6 large, juicy-ripe tomatoes, cut in ½-inch wedges
¼ cup chopped fresh basil, or 1 tablespoon dried basil
6 ears of corn (remove kernels with a sharp knife, working from top to bottom)

**1.** Heat the oil in a large skillet and sauté the onion and garlic for 5 minutes.

**2.** Add the tomatoes and basil, cover, and simmer for 15 minutes.

**3.** Stir in the corn kernels (about 3 cups), cover, and cook 5 minutes longer.

# SAUTÉED CORN AND MUSHROOMS

PREPARATION: 10 MINUTES          COOKING: 8–10 MINUTES          YIELD: 6–8 SERVINGS

2 tablespoons olive or vegetable oil
1 medium onion, chopped fine
1 clove garlic, minced or crushed (optional)
1 cup sliced mushrooms
3 cups corn kernels
¼ cup thinly sliced black olives
1 teaspoon dried thyme leaves
½ teaspoon coarsely ground black pepper

**1.** Heat the oil in a large skillet and sauté the onion for 3 minutes.

**2.** Add the garlic and mushrooms and sauté 2 minutes.

**3.** Stir in the corn, sliced olives, thyme, and black pepper. Cover and cook 5 minutes.

# CREAMED CORN

PREPARATION: 5 MINUTES          COOKING: 5 MINUTES          YIELD: 4 SERVINGS

*I'm not crazy about creamed vegetables, but creamed corn is something else. Some people make this dish with heavy cream or at least whole milk, whereas I'm satisfied with low-fat milk. Sometimes I spike this with a teaspoon of curry powder.*

2 tablespoons butter blend
2 tablespoons all-purpose flour or whole wheat flour
1 cup low-fat milk (whole milk or light cream)
2 cups corn kernels, or 4 ears of corn (working from top to bottom, remove kernels with a sharp knife)
¼ teaspoon coarsely ground black pepper

**1.** Melt the butter blend in a 2-quart saucepan.

**2.** Stir in the flour and cook for 1 minute.

**3.** Add the milk gradually, stirring constantly or whisking to eliminate any lumps from forming.

**4.** Stir in the corn (if using canned corn, substitute the corn liquid for an equal quantity of milk) and black pepper and cook gently for 2 minutes. Remove from the heat and serve.

# SUCCOTASH

*Succotash was the name the Algonquin tribes gave to their mixture of corn and lima beans. When the Narraganset Indians introduced this dish to the settlers in the 1620s, the name went with it. Squash would also be added to the pot, and, later on, onions and peppers were included. Bacon gave it fat and flavor.*

4 slices bacon
2 tablespoons butter blend
1 medium onion, chopped
2 cups corn kernels
2 cups cooked lima beans (if using fresh, cook 10 minutes in boiling water)
½ teaspoon coarsely ground black pepper
2 tablespoons chopped parsley

**1.** Cook the bacon in a large skillet, until crisp. Remove and drain on paper towels. Discard all the fat.

**2.** Melt the butter blend in the skillet and sauté the onion for 5 minutes.

**3.** Add the corn, cooked lima beans, and black pepper. Cover and cook for 5 minutes.

**4.** Sprinkle with chopped parsley and crumbled bacon before serving.

# CHILLED CORN SOUFFLÉ

PREPARATION: 10 MINUTES          CHILLING: 1½–2 HOURS          YIELD: 4 SERVINGS

*Serve this with watercress and hearty whole-grain bread, and you have a truly delicious supper.*

1 tablespoon butter blend or margarine
1 tablespoon all-purpose flour
1 cup low-fat milk
½ teaspoon coarsely ground black pepper
1 teaspoon curry
1 tablespoon gelatin granules (½ ounce)
1½ cups corn kernels
2 large eggs, separated

**1.** Melt the butter blend in a saucepan, stir in the flour, and cook 1 minute.

**2.** Add the milk slowly, stirring or whisking constantly to prevent lumping. Cook over low heat for 2 minutes.

**3.** Remove from the heat and add the pepper, curry, and gelatin. Stir to combine and allow to dissolve for 3–4 minutes.

**4.** Purée or mash 1 cup of the corn kernels and combine with the egg yolks and the remaining whole corn. Mix with the gelatin mixture.

**5.** Beat the egg whites until stiff and stir one-third into the corn mixture. Fold the rest of the whites in very gently.

**6.** Spoon into a ½–2-quart soufflé dish and refrigerate for about 4 hours.

# CASSEROLES

CHAPTER 5

*The little cares that fretted me,*
*I lost them yesterday. . . .*
*Among the husking of the corn*
*Where drowsy poppies nod*
*Where ill thoughts die and good are born*
*Out in the fields with God.*

—ANONYMOUS, 1898

**C**asserole dishes in general tend to be rather down-home affairs and for that reason they can be considered "comfort" foods. The recipes falling under that category in this cookbook include soufflés, custard puddings, and spoonbread. They don't take too much time to assemble and bake, and the result is usually a satisfying, stick-to-the-ribs dish, especially wholesome because it contains cornmeal, corn bread, or whole corn.

Some baked dishes, like the quiche and the corn stratas, are hearty enough to be served as a main meal. The soufflés, spoonbread, and puddings are perfect for a lunch or light dinner and as side dishes to accompany meat and poultry entrées.

Spoonbread, which originated in the South during colonial times, is neither bread nor soufflé but has a consistency somewhere in between. Although the eggs are separated, the cornmeal gives the batter a denser texture than a regular soufflé, but it has to be spooned onto a dish because it is not firm enough to slice like corn bread. Some cooks add a cup of slivered ham to the recipe; others serve it plain and simple and sometimes drizzled with honey or maple syrup for breakfast. Whichever way you serve it, it's sure to please. Without the addition of the corn kernels, it is my favorite form of corn "bread" which, if it lasts long enough, can be sliced when cold.

# CORN FRITTATA

PREPARATION: 15 MINUTES      BAKING: 15 MINUTES      YIELD: 4 SERVINGS

*Frittatas are so versatile. You can make them for yourself from leftover vegetables or fruit. I do this about once a week to use up odds and ends and make a satisfying lunch at the same time. I sometimes ask a friend in to join me.*

*You can also plan ahead and make a frittata from something sublime. Then it turns into a dinnertime meal for the family. Accompany this one with homemade muffins and a mixed green salad.*

3 tablespoons vegetable oil
1 small onion, chopped
1 clove garlic, crushed or minced
1 small zucchini, sliced
1 small red bell pepper, diced
1½ cups corn kernels (scraped from 3 ears of fresh corn)
2 tablespoons fresh chopped basil, or 1 teaspoon dried basil
½ teaspoon coarsely ground black pepper
8 large eggs
¼ cup low-fat milk, vegetable juice, or water

**1.** Preheat the oven to 350°F.

**2.** Heat 2 tablespoons of the vegetable oil in a 10-inch or thereabouts ovenproof skillet.

**3.** Over medium heat, sauté the onion, garlic, zucchini, and red pepper for 5 minutes.

**4.** Add the corn, basil, and black pepper and cook 2 minutes.

**5.** Blend or beat together the eggs and milk. Pour over the vegetables and cover the skillet.

**6.** Place in the oven and bake 15 minutes, or until the eggs are cooked to your liking. Serve with chopped tomatoes and scallions.

# CORN SOUFFLÉ

*To make this into an extra-special, eye-pleasing lunch or dinner, sprinkle with 4 slices of crispy, crumbled bacon and serve with a little fresh tomato sauce or fresh basil (pesto) sauce.*

2 tablespoons olive or vegetable oil
3 tablespoons all-purpose flour
⅛ teaspoon ground nutmeg or mace
¼ teaspoon ground black pepper
1 cup low-fat milk
½ cup cottage cheese
½ cup finely grated Parmesan cheese
3 egg yolks
1½ cups corn kernels (if using canned, substitute ¼ cup liquid for ¼ cup of the milk)
4 egg whites

**1.** Preheat oven to 375°F. and grease a 1½ quart, 3-inch-deep soufflé or other dish.

**2.** Blend or beat the first seven ingredients together.

**3.** Pour into a saucepan and heat gently for 4 minutes, stirring frequently.

**4.** Remove from the heat and stir in the egg yolks.

**5.** Add the corn kernels.

**6.** Beat the egg whites in a large bowl until stiff. Pour one-third of the corn mixture into the beaten whites and blend.

**7.** Fold in the rest of the cooked mixture gently just until a lumpy batter is formed.

**8.** Pour into the prepared dish and cut a circle in the top with a knife about 1 inch from the edge.

**9.** Bake in the hot oven for 30 minutes. Remove and serve at once.

# CORNMEAL CHEESE SOUFFLÉ

PREPARATION: 15 MINUTES     BAKING: 35 MINUTES     YIELD: 4 SERVINGS

*Try this soufflé served with jalapeño sauce, available where Mexican food is sold.*

2 cups low-fat milk
½ cup cornmeal or quick-cooking grits
3 tablespoons butter blend or margarine
¾ cup grated jalapeño or plain Jack cheese
    with ¼ teaspoon ground black pepper
    and ½ teaspoon ground cumin
3 eggs, separated

**1.** Preheat the oven to 350°F. and grease a 1½-quart soufflé or straight-sided casserole.

**2.** In a double boiler set over water, heat the milk until almost boiling.

**3.** Stir in the cornmeal or grits and the butter and cook, stirring occasionally, for 10 minutes, or until the mixture becomes a thick batter.

**4.** Remove the double boiler from the heat and beat in the cheese and egg yolks.

**5.** Beat the egg whites in a large bowl until stiff. Fold the cornmeal mixture into the beaten egg whites and pour into the prepared dish.

**6.** Bake 35–40 minutes until puffed and golden brown. Serve immediately.

# GRITS AND CHEESE PUDDING

PREPARATION: 20 MINUTES          BAKING: 40 MINUTES          YIELD: 6–8 SERVINGS

5 cups water
1 cup grits
1 cup grated cheese (a mix of leftover
  pieces works well)
2 tablespoons butter blend
1 cup low-fat milk
2 large eggs
2 tablespoons snipped chives or scallion
  greens
½ teaspoon ground black pepper

**1.** Preheat oven to 325°F. and grease a 1½-quart casserole.

**2.** Bring the water to a boil in a 2-quart saucepan and slowly add the grits. Return to a boil, reduce the heat, and simmer for 15 minutes, stirring occasionally.

**3.** Remove from the heat and stir in the cheese and butter.

**4.** Beat together the milk, eggs, chives, and pepper.

**5.** Stir into the grits mixture and pour into the casserole.

**6.** Bake for 40–45 minutes.

# CORN CUSTARD PUDDING

PREPARATION: 15 MINUTES          BAKING: 30 MINUTES          YIELD: 4–6 SERVINGS

*This is like a deep quiche without the crust.*

2 tablespoons olive or vegetable oil
1 medium red onion, chopped
1 sweet red pepper, diced
1½ cups corn
3 eggs, beaten
½ cup low-fat milk
¾ cup grated cheddar cheese
½ teaspoon dried basil, or 1 tablespoon
   chopped fresh basil
¼ teaspoon ground black pepper

**1.** Preheat the oven to 350°F. and grease a 1½-quart casserole.

**2.** Heat the oil in a skillet and add the onion and red pepper. Sauté for 5 minutes.

**3.** Remove from the heat and place in a medium mixing bowl. Add the rest of the ingredients. (If using canned or frozen corn, drain off the liquid and use to replace some of the milk.)

**4.** Pour into the prepared casserole and bake for 30 minutes, or until a knife inserted in the center comes out clean.

# SKILLET CORN PANCAKE

PREPARATION: 15 MINUTES          BAKING: 25 MINUTES          YIELD: 4–6 SERVINGS

*You don't have to use a skillet for this. Substitute any shallow baking dish.*

6 slices bacon
2 tablespoons butter blend or margarine
1 cup low-fat milk
⅔ cup all-purpose flour
⅓ cup whole wheat flour
2 large eggs, separated
½ teaspoon ground black pepper
1 cup corn kernels (drained if canned or frozen)

**1.** Preheat oven to 450°F. and grease a 9-inch by 2-inch-deep dish (or use an ovenproof skillet).

**2.** Cook the bacon for 4 minutes, or until crisp. Remove from the skillet and drain on paper towels.

**3.** Pour off the bacon fat and put the butter blend in the ovenproof skillet or shallow dish. Place in the preheated oven.

**4.** Blend or process the milk, flours, egg yolks, pepper, and corn for 30–40 seconds.

**5.** Beat the egg whites until stiff and stir into the corn mixture. Add the crumbled bacon.

**6.** Pour into the hot skillet or dish and bake for 15 minutes. Reduce the heat to 400° and bake for 10 minutes longer. Best when served warm.

# CORNED BEEF AND CORN STRATA

PREPARATION: 15 MINUTES          BAKING: 40 MINUTES          YIELD: 6 SERVINGS

4 large corn muffins or 8 slices corn bread
1 cup grated cheddar cheese
1 12-ounce can corned beef, cut into 9 slices
1½ cups corn kernels
3 eggs
⅛ teaspoon each cayenne pepper and ground nutmeg
1½ cups low-fat milk
¼ cup finely grated Parmesan cheese

**1.** Preheat the oven to 350°F. and grease a 2-quart soufflé or straight-sided deep ovenproof dish.

**2.** Cut the muffins or pieces of corn bread into ½-inch-thick slices (you should have approximately 16 slices) and make a tight layer on the bottom of the greased dish.

**3.** Cover with ⅓ cup cheddar cheese followed by 3 slices corned beef and ½ cup corn.

**4.** Repeat these layers two more times and top with the remaining slices of corn bread.

**5.** Beat together the eggs, seasonings, and milk. Pour over the layers.

**6.** Sprinkle with the Parmesan cheese and bake for 40 minutes.

# MARTHA MOLNAR'S CORN AND GREEN PEPPER PUDDING

PREPARATION: 10 MINUTES          BAKING: 40 MINUTES          YIELD: 4 SERVINGS

*I knew Martha from my days as a reporter at the Patent Trader newspaper in Westchester, New York. We were always trading recipes that could be assembled in minutes when we arrived home after 6 P.M. This one goes together very quickly, especially if you cooked the bacon the night before. Martha sometimes uses 1 cup heavy cream where I have used all low-fat milk.*

6–8 slices bacon
1 small onion, chopped
1 small green bell pepper, chopped
2 cups corn kernels
1 cup fresh or packaged breadcrumbs
2 large eggs
2 cups low-fat milk
¼ teaspoon ground black pepper

**1.** Preheat the oven to 375°F. and grease a 1½–2-quart casserole.

**2.** Fry the bacon, remove, and reserve 2 tablespoons of the fat in the skillet.

**3.** Sauté the onion and green pepper in the bacon fat for 2 minutes.

**4.** Stir in the corn and breadcrumbs. Remove from the heat.

**5.** Beat the eggs, milk, and pepper together and add the breadcrumb mixture. Combine and pour into the casserole.

**6.** Bake for 40 minutes.

# MEXICAN STRATA

PREPARATION: 10 MINUTES          BAKING: 40 MINUTES          YIELD: 6 SERVINGS

*Serve with chopped jalapeño peppers, sour cream, and yellow rice. (Add ½–1 teaspoon turmeric to the water in which the rice is cooked.)*

9 flat corn tostadas
1½ cups kidney beans
1½ cups corn kernels
1½ cups grated plain or jalapeño Jack
  cheese
1 small onion, chopped
1 small green bell pepper, chopped small
2 cups crushed tomatoes
2 cloves garlic, crushed
½ teaspoon ground cumin
½ teaspoon chili powder

**I.** Preheat the oven to 350°F. and grease a 2-quart soufflé dish or casserole.

**2.** Break the tostada shells in half and arrange 6 pieces on the bottom of the dish.

**3.** Layer ½ cup each of kidney beans, corn, and cheese. Sprinkle with onion and green pepper. Repeat the layers two more times.

**4.** Combine the tomatoes, garlic, cumin, and chili powder. Pour over the last cheese layer.

**5.** Place in the oven and bake for 40 minutes.

# CHEESE, MACARONI, AND CORN CASSEROLE

PREPARATION: 15 MINUTES          BAKING : 25 MINUTES          YIELD: 4 SERVINGS

*This is one of our family favorites.*

8 ounces elbows, twists, or small shells
  (4 cups cooked)
2 tablespoons butter blend or margarine
2 tablespoons all-purpose flour
2 cups low-fat milk
1¼ cups grated cheddar cheese
½ teaspoon ground black pepper
¼ teaspoon ground nutmeg
1½ cups corn kernels
2 tablespoons grated Parmesan cheese
2 tablespoons fresh breadcrumbs

**1.** Preheat oven to 350°F. and grease a 2-quart ovenproof casserole dish.

**2.** Cook the macaroni until al dente, about 8 minutes.

**3.** Meanwhile, melt the butter blend in a saucepan and stir in the flour. Cook for 1 minute.

**4.** Add the milk and whisk until the sauce is smooth. (If using canned corn, substitute the liquid for some of the milk.)

**5.** Remove from the heat and stir in the cheese and seasonings.

**6.** Drain the macaroni after about 8 minutes cooking time and combine with the corn into the cheese sauce. Pour into the casserole dish.

**7.** Sprinkle the top with the Parmesan and breadcrumbs and bake for 25 minutes.

# CORN QUICHE

PREPARATION: 20 MINUTES          BAKING: 45 MINUTES          YIELD: 4–6 SERVINGS

## CRUST

1 stick butter blend (or ½ cup solid
    shortening)
1 cup all-purpose flour
⅓ cup whole wheat flour
2–3 tablespoons ice-cold water
1 egg white, beaten (to be reserved)

**1.** Place the flours in a large bowl.

**2.** Cut the butter blend into small pieces and mix into the flours until the mixture resembles peas.

**3.** Add the cold water, 1 tablespoon at a time, until a ball forms.

**4.** Place the ball of dough onto a sheet of wax paper, press into a circle, cover, and refrigerate.

## FILLING

1 tablespoon olive or vegetable oil
1 small onion or 4 shallots, chopped
3 large eggs
1½ cups low-fat milk
1 cup cottage cheese
½ cup grated cheddar cheese
1½ cups corn kernels
¼ teaspoon each ground black pepper,
    nutmeg, and mace

**1.** Preheat the oven to 425°F. and grease a 9–10-inch quiche dish (or use a 2-inch-deep pie dish).

**2.** Heat the oil moderately in a skillet and sauté the chopped onion for 2 minutes. Place in a large mixing bowl.

**3.** Add the eggs, milk, and cottage cheese and beat together.

**4.** Stir in the grated cheddar, the corn kernels (if using canned or frozen corn, drain thoroughly), and the seasonings.

**5.** Roll out the pastry to about ⅛-inch thickness. Fit into the pie plate, trim the pastry ½ inch above the rim and flute the edges. Brush the bottom and sides with the egg white.

**6.** Pour the filling into the pastry and bake 10 minutes. Reduce the heat to 350° and continue baking for 30 minutes more. A knife inserted into the center should come out clean.

**7.** Allow to cool a while before cutting into pieces.

# CORNMEAL POPOVERS

| PREPARATION: 5 MINUTES | BAKING: 25 MINUTES | YIELD: 12 POPOVERS |
| --- | --- | --- |

*Popovers always seem to be a real treat. These are particularly easy and don't take too long to bake.*

¼ cup vegetable oil
½ cup cornmeal
½ cup all-purpose flour
1⅓ cups low-fat milk
2 large eggs, beaten

**1.** Preheat the oven to 425°F. and grease 12 6-ounce custard or muffin cups. Drop 1 teaspoon vegetable oil into the bottom of each cup. Place the cups on a tray and preheat in the oven.

**2.** Beat the cornmeal, flour, milk, and eggs together until smooth.

**3.** Remove the custard cups from the oven. Half fill with the batter.

**4.** Bake for 30 minutes and serve hot.

# WHOLE CORN SPOONBREAD

PREPARATION: 15 MINUTES     BAKING: 45 MINUTES     YIELD: 6–8 SERVINGS AS A SIDE DISH

*Contrary to its name, spoonbread is actually a pudding, or a soufflé. Serve it as a side dish in place of rice or pasta. Or turn it into a main meal accompanied by a vegetable and salad. Vary the flavors by adding 1 cup sautéed chopped onion or 1 cup grated cheddar cheese. You could even add 1 teaspoon of your favorite herb.*

*There's no reason that you can't make it into a sweet dessert; add ¼ cup sugar plus 1 cup berries, chopped fruit, or nuts.*

1½ cups low-fat milk
1 cup buttermilk
1 cup cornmeal
1 cup corn kernels
¼ cup unsalted butter blend or margarine
1 tablespoon sugar
½ teaspoon baking soda
3 large eggs, separated

**1.** Preheat the oven to 350°F. and grease a 1½–2-quart casserole.

**2.** Heat the milk and buttermilk in a large saucepan.

**3.** When it begins to boil, add the cornmeal slowly in a thin stream, stirring until the mixture is thick and smooth.

**4.** Remove from the heat and stir in the corn kernels, butter, sugar, and baking soda.

**5.** Beat the egg yolks and stir into the corn mixture.

**6.** Beat the egg whites in a large mixing bowl until stiff and gently fold into the batter.

**7.** Pour into the greased casserole and bake for 45 minutes. The spoonbread will be golden brown and puffed like a soufflé. Serve immediately.

# ETHNIC SPECIALTIES

C H A P T E R   6

## PENNSYLVANIA DUTCH DRIED CORN

The old method of drying corn is to cut the kernels from the cobs and sprinkle in a single layer on a baking tray. Place in 250°F oven to dry slowly for two days. The kernels must be stirred around occasionally. After this time, put the dried kernels into a cloth bag and hang to dry in a warm place. When dry, the corn is ready to be stored in a jar for future use.

Dried corn can be reconstituted by soaking overnight in water (1 cup corn to 2 cups water) and simmering it for 1 hour. Add a little butter, pepper, and cream to taste. A cup of dried corn makes 6 cups.

**A**lthough corn is grown and eaten all over the world today, if you ask the average American to comment on ethnic dishes made with or from corn, he would undoubtedly think of Mexican tortillas, the flat breads made from *masa harina*. (Finer than regular cornmeal, masa harina is cornmeal flour made from cooked and dried corn, as opposed to uncooked dried corn.) Today, tortillas are popular all over the United States. We eat soft tortillas stuffed and rolled for enchiladas, flautas, and quesadillas, and crisp, fried tortillas for tacos, tostadas, and nachos.

In the Southwest, corn is a major ingredient in Tex-Mex cuisine, and Native Americans eat it fresh, dried, and ground as a staple of the diet. As in Central America, the peoples of South America eat it fresh or dried in many delicious ways. The Pennsylvania Dutch make corn into wonderful relishes and dry quantities for their stewed dried corn dishes.

Corn has been grown in Europe since the sixteenth century when explorers took it back with them from America. Grown first by the Spanish, it didn't take long to reach Turkey and Italy where it was called tasseled "wheat." Corn was particularly favored by the Italian farmers. Ask an American-Italian to describe a dish made from cornmeal and he will immediately wax eloquent on *polenta* (a variation of American cornmeal mush) and corn-meal gnocchi (tiny dumplings). Those of Rumanian descent call their mush *mamaliga*, which, like polenta, is eaten as a staple like bread, pasta, and potatoes.

Here, then, is a sprinkling of ethnic dishes.

# CORN TORTILLAS

PREPARATION: 30 MINUTES          COOKING: 2 MINUTES EACH          YIELD: 12 TORTILLAS

*Although a tortilla press makes perfect flat rounds, it's not worth purchasing one unless you plan to make hundreds. This hand-rolled method is quite easy. (If you have a 13-inch skillet, you can cook 4–5 at one time.)*

**2 cups masa harina (corn flour)**
**1¼ cups warm water**

**1.** In a medium bowl, combine the masa harina and water.

**2.** When the dough begins to form a ball, use your hands to make it smooth.

**3.** Divide the dough into 12 equal portions and form them into balls.

**4.** Place each ball between two sheets of wax paper.

**5.** Flatten and roll into a 6-inch circle; leave between the wax paper.

**6.** Continue until you have a stack of 12 tortillas.

**7.** Heat a large, heavy (cast-iron) skillet and remove several tortillas from their wax papers. Without greasing the skillet, fry the tortillas about 1 minute each side. They should be soft. Keep warm (and soft) by enveloping them in aluminum foil.

**8.** Repeat with the remaining tortillas.

**9.** Serve as bread, or stuff for enchiladas, flautas, and tacos.

# TACO SHELLS

*You can stuff tacos (or pile tostadas—see Variation) with cooked ground meat, heated leftover chopped chicken, crushed kidney or pinto beans left cold or heated up, even cold, canned, flaked tuna. Drizzle with a spicy Mexican sauce, then start adding layers of grated cheddar or Jack cheese, chopped tomatoes, shredded lettuce, and sliced onions. This is a really well-rounded meal, one that I love my nine-year-old daughter Wendy to take packed into her lunch box. She has to assemble it herself at school, of course.*

**¼ cup vegetable oil**
**corn tortillas**

**1.** Using a heavy skillet, heat the oil over medium.

**2.** Place 1 tortilla at a time into the oil and fry for 20 seconds.

**3.** Remove from the oil with blunt-tipped tongs and fold the tortilla over in half to make a taco shape.

**4.** Return to the hot oil and fry for 30 seconds, keeping the edges apart with the aid of the tongs.

**5.** Turn and fry until crisp, about 30 seconds. Remove from the oil with tongs and drain on paper towels.

**6.** Repeat with remaining tortillas.

## TOSTADAS AND NACHOS

*To make **tostadas**, place tortillas in a single layer in the hot oil and fry for 1½ minutes. Remove with tongs and drain on paper towels.*

*To make **nacho chips**, cut the tortillas into quarters and follow the instructions for tostadas. Spread the chips with mashed kidney beans, sprinkle with grated cheese, and top with a slice of hot jalapeño pepper or a dab of spicy Mexican sauce. Pop under the broiler for a minute or two until the cheese melts, microwave for 50 seconds on high, or place on a cookie sheet in a 350°F. oven for 5 minutes.*

# CHICKEN ENCHILADAS

PREPARATION: 25 MINUTES     COOKING/BAKING: 15 MINUTES     YIELD: 4 SERVINGS

*If you don't want to make your own soft tortillas, buy them at the supermarket or use the recipe for plain cornmeal pancakes on page 138.*

*I serve this for dinner with crushed cooked kidney beans heated and mixed with ½ cup sliced scallions (you could also add some crumbled bacon) and yellow rice (made with brown rice and 1 teaspoon turmeric).*

1 pound boneless chicken breasts
¼ cup chicken stock, tomato juice, or water
6 tablespoons cream cheese
2 tablespoons chunky Mexican sauce (mild or hot)
1 tablespoon vegetable oil
1 small onion, chopped
3 cloves garlic, crushed or minced
1 28-ounce can crushed tomatoes (3 cups)
1 3½-ounce can green chili peppers, chopped
1 teaspoon ground coriander
¼ cup vegetable oil
8 6-inch tortillas
1½ cups sour cream
¼ cup chopped chives or scallion greens

**1.** Preheat the oven to 400° F.

**2.** Put the chicken breasts in a skillet, pour in the stock, cover and simmer for 10 minutes.

**3.** Reserve ⅓ cup of the liquid. Thinly slice the chicken breast.

**4.** Blend the cream cheese, reserved liquid, and Mexican sauce together. Stir in the chicken strips.

**5.** While the chicken is poaching, heat the 1 tablespoon oil in a large skillet and sauté the onion and garlic for 2 minutes.

**6.** Add the tomatoes, chili peppers, and coriander. Simmer for 15 minutes.

**7.** Heat the ¼ cup oil in a skillet and drop each tortilla into the hot fat for about 10 seconds. This is not to cook them but to make them pliable.

**8.** Remove immediately to a baking dish measuring approximately 12″ × 9″ × 2″ and spoon 2–3 tablespoons of the chicken mixture along the center of each tortilla. Roll up and place seam-side down in the dish. Repeat until all the tortillas and the filling have been used up.

**9.** Spoon the tomato sauce over the top and bake for 15 minutes.

**10.** Mix the sour cream and chives and spoon over each cooked tortilla.

# RUMANIAN MAMALIGA

COOKING: 15 MINUTES                    YIELD: 4 SERVINGS

*Another version of cornmeal porridge. Like regular porridge this is served hot. Top with butter, yogurt, or sour cream. Like polenta, this can be spread into a flat-bottomed dish and allowed to cool, then cut into slices and fried in hot butter 1–2 minutes a side.*

**3 cups water**
**1 cup coarse cornmeal**

**1.** Bring the water to a boil in a saucepan and slowly add the cornmeal in a thin stream, stirring constantly.

**2.** When the mixture is smooth, cover and simmer for 10 minutes.

# MEXICAN CORN HUSK TAMALES

PREPARATION: 1½–2 HOURS          COOKING: 45 MINUTES          YIELD: 15–20 TAMALES

*Tamales were made by the Aztec Indians in both savory and sweet forms. This recipe is savory (and very simple). The fresh corn mixture can be replaced with 3 cups of cooked spicy ground beef mixture. If there's no fresh corn available, you can buy packets of corn husks (and also the masa harina flour) at markets specializing in Hispanic products. Purchased husks are easier to work with because they are all the same size. The husks will need soaking in warm water for 1½ to 2 hours before you use them, so take this into consideration when you plan your recipe schedule. You can substitute canned or thawed frozen corn kernels, if necessary. Tamales are also served as dessert. For sweet tamales, use 1½ cups puréed fruit combined with ¼ cup sugar, or to taste. Another filling can be made of preserves combined with shredded coconut or ground nuts.*

6 large ears of corn (you need 15–20 wrappers and 2 cups corn)
2 cups masa harina flour
1¼ cups water
¾ cup soft shortening (I use margarine—authentic recipes use lard)
1 large sweet chili pepper (those used for chile rellenos), chopped fine
1 small onion, grated
½ cup grated Jack cheese
½ cup chunky taco sauce (medium hot or mild)

**1.** Shuck the corn a day ahead of time so that the husks can dry. Remove the kernels from the cobs and refrigerate. You will have about 3 cups of corn but for this recipe you need only 2 cups.

**2.** Soak the corn husk wrappers in warm water 1–2 hours ahead of time.

**3.** Beat together the flour, water, and shortening in a medium bowl. This is the masa dough. Cover while you prepare the filling.

**4.** I like to use a food processor and quickly process the chili pepper and onion. I throw in the corn for about 10 seconds. If you don't have a processor, place the 2 cups corn kernels in a medium bowl and add the chopped chili pepper, grated on-

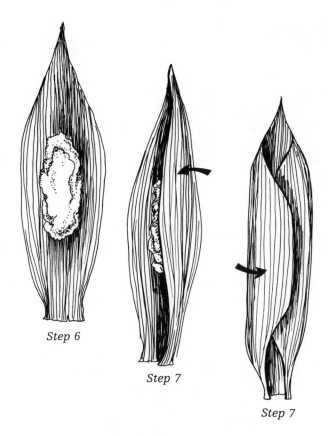

Step 6

Step 7

Step 7

Step 8

ion, grated cheese, and taco sauce. Stir together.

**5.** Pat dry the corn husk wrappers. Place the husks with the tip facing away from you.

**6.** Spoon 2–3 tablespoons dough onto the wrapper and spread into a rectangle that reaches to the right edge of the husk and 2–3 inches from the top and bottom with a 1-inch margin of husk on the left.

**7.** Spoon 1½–2 tablespoons filling into the center of the dough and fold the right side of the husk over to the center of the filling. Fold the left side over the dough.

**8.** Fold the bottom and top of the husk to meet in the center. Continue to assemble the tamales until the masa dough and the fillings have been used up.

**9.** Place the tamales, seam-side down, on a rack over 1 inch of boiling water in a pan. Cover and steam for 45 minutes or until a husk peels easily away from the dough.

**10.** Remove the husks to serve, if desired, and pass a dish of hot taco sauce.

# BASIC POLENTA

COOKING: 20 MINUTES                    YIELD: 6 CUPS

*Living in Geneva, Switzerland, for seven years during the 1960s was my passport to sampling many European cuisines. Northern Italian food became my passion, and polenta, a ridiculously simple dish of cooked cornmeal, numbered among my favorite recipes. Although sometimes eaten as an appetizer topped with mushrooms or melted gorgonzola, it was more often served as a side dish to accompany chicken or game. On those occasions, it would be covered in a rich brown gravy or a tomato or vegetable sauce.*

*There are many non-traditional ways to add flavor and nutrition. Make polenta for brunch and top it with fried eggs and taco sauce; sprinkle it with grated cheddar cheese and chopped onions; top with slices of mozzarella and tomato sauce and pop under the broiler. You can chill polenta for about 2 hours, cut into slices and fry in a little butter or olive oil until browned on both sides. Serve with honey, preserves, tomato sauce, or any gravy. With such a bland base, anything goes.*

*The best polenta is made from stoneground or coarse ground yellow cornmeal. Although it is possible to find imported "polenta" cornmeal in some specialty shops and Italian stores, good old American cornmeal works just as well.*

**7 cups water**
**2 cups coarse cornmeal**

**1.** Bring the water to a rolling boil in a large pot.

**2.** Reduce the heat and pour in the cornmeal in a thin, slow stream, stirring constantly and rapidly with a wooden spoon. (This is the way we make Scotch porridge—stop stirring or hurry the pouring and you end up with a lumpy mess.)

**3.** Continue stirring for about 5 minutes, lower the heat, and simmer for 15 minutes, still with an occasional stir.

**4.** Serve hot with a sauce or spoon into a flat-bottomed dish and chill.

# POLENTA GNOCCHI

PREPARATION: 30 MINUTES      CHILLING: 2 HOURS      YIELD: 4–6 SERVINGS
COOKING: 5 MINUTES PER BATCH OF 12

*Serve these with a hearty chicken stew. You can also arrange the gnocchi on a greased dish, cover with a vegetable sauce, and bake at 350°F. for 20 minutes.*

4 cups water
1 cup cornmeal
2 tablespoons olive oil
1 large egg, beaten
¼ cup grated Parmesan cheese
½ teaspoon coarsely grated black pepper
¼ cup vegetable oil

**1.** Bring the water to a rolling boil and stir in the cornmeal in one continuous slow stream. Continue stirring for 5 minutes.

**2.** Simmer and stir frequently for 15 minutes longer.

**3.** Remove from the heat and beat in the oil, egg, cheese, and pepper.

**4.** Spoon the mixture into a flat-bottomed rectangular or square dish and smooth down until it is about 1 inch thick. Refrigerate for 2 hours.

**5.** Slice into 1-inch pieces.

**6.** Heat 2 tablespoons of the vegetable oil in a large, heavy skillet until medium-hot and cook half the gnocchi at a time, turning once after 1–2 minutes. Fry until they are golden brown. Remove to a plate lined with paper towels and place in a low oven to keep warm.

**7.** Repeat with the remaining gnocchi.

# ARGENTINE PUCHERO

PREPARATION: 30 MINUTES          COOKING: 60 MINUTES          YIELD: 6–8 SERVINGS

*This is what we call "hotpot" in the British Isles, and a very delicious stew it is.*

¼ cup olive oil
1 cut-up 3½–4-pound chicken, with skin removed
1 pound lean, boneless lamb, cut in 2-inch cubes
4 medium onions, sliced very thin
5 cloves garlic, crushed or minced
1 large carrot, sliced
2 zucchinis or ½ a small (around 3 pounds) pumpkin (peeled), cut in half-inch slices or cubes
1 tablespoon dried thyme
⅛ teaspoon cayenne pepper
1 cup red wine
2 cups chicken stock
4 cups whole tomatoes with juice, roughly cut up
3 cups corn kernels
3 tablespoons cornstarch
2 tablespoons water

**1.** Heat 2 tablespoons of the oil in a 5–6-quart Dutch oven and sauté the chicken pieces 1½ minutes on each side. Remove to a plate.

**2.** Add the lamb pieces to the pot and brown on all sides for a total of 2 minutes. Remove to a plate.

**3.** Heat the remaining 2 tablespoons of oil in the pot and sauté the onions, garlic, and carrot for 5 minutes.

**4.** Add the squash or pumpkin pieces, stir, and cook for 3 minutes.

**5.** Sprinkle with thyme and cayenne and cover with the wine, stock, and tomatoes.

**6.** Return the chicken and lamb to the pot, combine, and partially cover the pot.

**7.** Cook slowly for 45 minutes.

**8.** Add the corn to the pot for the last 15 minutes.

**9.** Combine the cornstarch and water to make a smooth paste and stir into the hot liquid for the last 5 minutes to thicken the stew.

# FISH POSOLE

PREPARATION: 20 MINUTES       COOKING: 20 MINUTES       YIELD: 4–6 SERVINGS

*In the Southwest, posole is a stew made with hominy (corn kernels that have been dried, then soaked in a lime solution to remove the skins). The Aztec Indians called hominy* nixtamal *and the method of preparing it and then boiling the kernels to soften them,* pozole. *Softened pozole (or posole) was then added to the stew pot. Cans of hominy can be purchased at supermarkets and stores in ethnic neighborhoods. On occasion, I have used regular canned corn.*

*Here, then, is a fish posole—or make it with chicken or pork if you prefer; just remember to increase the cooking time by about 30 minutes and add 1 teaspoon of chili powder. Try this recipe served with soft, warm tortillas (see page 76), mashed avocado with lime juice, chopped onion, and sour cream.*

1 tablespoon vegetable oil
1 medium red Bermuda onion, chopped fine
2–4 cloves garlic, crushed or minced
8 cups chicken stock or water
2 1-pound cans whole hominy, yellow or white
1 3½-ounce can mild green chili peppers, chopped
½ teaspoon cumin powder
¼ teaspoon dried hot chili pepper flakes
1 pound monkfish or cod fillet, cut into 1-inch pieces
1¼ pounds medium shrimp, shelled and cleaned

**1.** Heat the oil in a 5–6-quart Dutch oven and sauté the onion and garlic for 5 minutes.

**2.** Add the stock, hominy, chili peppers, cumin, and chili flakes.

**3.** Bring to a boil, reduce the heat, and simmer partially covered for 10 minutes.

**4.** Add the fish and shrimp and simmer for 5 minutes only.

**5.** Remove from heat and serve.

# MARTHA MOLNAR'S RUMANIAN CORNMEAL PUDDING

PREPARATION: 20 MINUTES          BAKING: 15 MINUTES          YIELD: 4 SERVINGS

*One of my reporter colleagues, Martha Molnar, is Rumanian. She passed along this recipe that she learned in her mother's home. This can be served with sour cream if you like. In fact, Martha likes to spread a layer of sour cream over each cheese layer before it is baked. Sometimes she uses layers of pepperoni slices instead of the bacon.*

1 recipe of Rumanian Mamaliga (p. 79)
   spread in a flat-bottomed dish and
   chilled
6 slices bacon
6 slices Swiss cheese
2 tablespoons melted butter
¼ cup grated Parmesan cheese
sour cream (optional)

**1.** Preheat the oven to 375°F. and grease a 1½-quart casserole dish.

**2.** Cut the mamaliga into 3″ × 2″ strips and cut those into ½-inch-thick slices.

**3.** Heat a skillet and cook the bacon until crisp. Drain on paper towels.

**4.** Put one layer (about one fourth) of the mamaliga in the casserole dish and crumble 2 slices bacon over the top.

**5.** Cover with 2 slices Swiss cheese. Repeat the layers two more times and end with a fourth layer of mamaliga.

**6.** Drizzle with the melted butter and sprinkle with the Parmesan cheese.

**7.** Bake for 15 minutes on the top shelf of the oven.

# MEAT, POULTRY, AND FISH

## CHAPTER 7

*I'm Captain Jinks of the Horse Marines,*
*I give my horse good corn and beans;*
*Of course 'tis quite beyond my means,*
*Though a Captain in the army.*

—T. MACLAGAN
CAPTAIN JINKS (1870)

I love to add corn to many of my dishes if it's only for the splash of color, especially when you mix it with red bell peppers and green zucchini or green scallion tops. But I also love the delicate, sweet flavor and the crunchy texture.

You can toss a cup or so of whole kernels into practically any dish—doesn't matter whether it's meat, fish, or chicken—without changing the flavor or liquid content. But you will increase the nutritional quality and have more servings to go around.

If you grind whole kernels or scrape them fresh off the cob (slice through the centers before scraping them off) you can mix them into meatballs, meatloaf, and dumplings. This makes for a closer texture than when you use whole kernels.

And then there's cornmeal batter. It's like having your bread and eating it when you top a stew with dumplings or corn bread. It's much more substantial than a feathery pastry pie crust. Somehow it's just better, more soul-satisfying, when you have corn bread moistened with all that good gravy. You don't have to stick with the plain corn bread topping, either. Add some chili peppers and some grated cheese and bake it over a pot of kidney beans and meat (as in chili con carne) and you'll have an easy dish to serve up for football-Sunday dinners.

# CHICKEN CORN PIE

4 tablespoons olive or vegetable oil
1 large onion, sliced thin
1 stalk celery, sliced thin
1 carrot, sliced thin
2–4 cloves garlic, minced or crushed
1 4-pound chicken, skinned and cut into
    serving pieces
3 tablespoons all-purpose flour
1 teaspoon coriander
1 teaspoon cumin
½ teaspoon coarsely ground black pepper
¼ teaspoon cayenne pepper (optional)
1½ cups chicken stock, tomato, or vege-
    table juice
1 recipe of corn bread batter (see recipe
    for Regular Corn Bread on page 114)

**1.** Preheat the oven to 425°F.

**2.** Heat 1 tablespoon oil in a large skillet and add the onion, celery, carrot, and garlic.

**3.** Sauté over a gentle heat for 3 minutes. Remove to a container.

**4.** Roll half of the chicken pieces in half of the flour and heat half of the remaining oil. Sauté over medium heat, turning to brown each side of the chicken, about 4 minutes. Remove the pieces to the vegetable dish.

**5.** Repeat with the rest of the oil, chicken pieces, and flour.

**6.** Return the sautéed chicken and vegetables to the skillet, add the seasonings and the chicken stock.

**7.** Stirring, cook over medium heat until the sauce thickens, about 2 minutes.

**8.** Pour into a deep 2½–3-quart baking dish and top with the corn bread batter.

**9.** Bake for 25 minutes.

# MONKFISH PRIMAVERA

PREPARATION: 10 MINUTES       COOKING: 12 MINUTES       YIELD: 4 SERVINGS

*With its red, green, and yellow colors, this is a very festive-looking dish. Monkfish has a chunky texture, something like deep sea scallops or chunks of chicken breast. I sometimes interchange them. You could also substitute shrimp. I like to vary the flavor by using 1 teaspoon of fresh chopped ginger root in place of the tarragon. I serve this with brown rice and curry mayonnaise.*

2 tablespoons, vegetable oil

2 cloves garlic, minced or crushed

2 small red bell peppers, cut in thin slivers

2 pounds monkfish (or 2 whole chicken breasts, boned), cut into 1-inch cubes

1 teaspoon dried tarragon, or 2 tablespoons fresh, chopped tarragon

½ teaspoon coarsely ground black pepper

2 cups young sugar snap peas, no more than 2 inches long

1 cup corn kernels

2 tablespoons chopped chives or scallion greens

**1.** Heat the oil in a large skillet and add the garlic and red peppers.

**2.** Sauté over medium heat for 2 minutes.

**3.** Add the cubed monkfish or chicken breasts and sprinkle with the tarragon and black pepper. Stir around in the skillet and cook for 5 minutes. (If substituting scallops or shrimp, cook for 1 minute only.)

**4.** Stir in the snap peas and corn, cover the skillets, and cook for 5 minutes longer. (Only 2–3 minutes if cooking scallops or shrimp.)

**5.** Arrange on plates and sprinkle with the chives.

## Curry Mayonnaise

*Just blend all the ingredients together.*

1 cup homemade or prepared mayonnaise

1 teaspoon curry powder

¼ teaspoon ground ginger

1 tablespoon lime or lemon juice

1 tablespoon honey

# CODFISH GUMBO

PREPARATION: 10 MINUTES          COOKING: 20 MINUTES          YIELD: 4 SERVINGS

*Halibut, monkfish, scallops, and shrimp can be substituted for the cod.*

2 tablespoons olive or vegetable oil
½ pound fresh okra, cut in ¼-inch slices,
 or a 10-ounce package frozen okra
1 medium onion, sliced very thin
2 cloves garlic, minced or crushed
1 8-inch zucchini, sliced thin
6 large tomatoes, chopped, or 1 29-ounce
 can crushed tomatoes
3 ears of corn, or 1½ cups canned or frozen kernels
¼ teaspoon black pepper
⅛ teaspoon cayenne pepper or crushed
 red pepper flakes
4 cod steaks (1½ pounds)

**1.** Heat the olive oil in a large skillet and sauté the okra over medium heat for 2 minutes.

**2.** Add the onion and garlic and sauté for 5 minutes.

**3.** Add the zucchini and tomatoes and cook over a gentle heat for 5 minutes.

**4.** Using a sharp knife and working from top to bottom, remove the kernels from the ears of corn. Add to the skillet with the black and cayenne peppers. Combine.

**5.** Lay the cod steaks on top, spooning the mixture over and around them.

**6.** Cook gently for 5–10 minutes, until the fish is cooked through.

# CURRIED CHICKEN AND CORN IN PATTY SHELLS

PREPARATION AND COOKING: 20 MINUTES                    YIELD: 4 SERVINGS

*This is my "cop-out" dinner. It never fails to please and it takes no time at all to prepare from leftovers. I don't like canned creamed corn but make my own sauce and use canned, frozen, or fresh corn kernels. Of course, the best creamed corn is right off the cob when you cut through the center of the kernels and scrape them off with the milk. But if you don't have fresh corn, the quick creamed corn is still very good. Serve with a big green salad or steamed broccoli.*

1 package frozen patty shells (6 shells)
3 cups creamed corn
2 cups cooked chicken, in ½-inch dice
1 teaspoon curry powder

**1.** Preheat the oven to 450°F. When it's hot, place the patty shells on a baking sheet, reduce the temperature to 425° and bake 20 minutes.

**2.** While the shells are baking, combine the creamed corn, chicken, and curry. Heat gently for 10 minutes or until hot.

**3.** Remove the shells from the oven and scoop off the lids.

**4.** Spoon the corn mixture into the patty shells.

## Creamed Corn

*To make your own see page 54, or cook the following in 3 minutes:*

2 tablespoons butter blend or margarine
2 tablespoons whole wheat or all-purpose flour
1 cup low-fat milk
¼ cup cream
2 cups corn kernels

**1.** Melt the butter blend in a saucepan over low heat and stir in the flour. Cook for 1 minute.

**2.** Pour in the milk and cream and stir or whisk until smooth, about 2 minutes.

**3.** Add the corn and stir to combine. Cook about 3 minutes, until the corn is heated.

# CORNMEAL DUMPLINGS

PREPARATION: 5 MINUTES          COOKING: 10 MINUTES          YIELD: 10 DUMPLINGS

*You can use any kind of ground meat for these. Beef and pork will make a moister dumpling, but I often use ground raw chicken or cooked ham.*

¼ pound pork
1 teaspoon dried thyme
¼ teaspoon coarsely dried black pepper
1 clove garlic, crushed
½ cup all-purpose flour
½ cup cornmeal
2 teaspoons baking powder
¼ cup low-fat milk or tomato juice (add
    ½ teaspoon baking soda if using juice)
1 large egg, beaten
simmering vegetable soup

**1.** Heat a skillet and sauté the pork, thyme, pepper, and garlic over medium heat for about 2 minutes.

**2.** Drain off the fat.

**3.** Sift the flour, cornmeal, and baking powder into a medium bowl and stir in the cooked pork.

**4.** Add the milk and beaten egg and stir until barely combined.

**5.** Using a large dessert spoon, drop the mixture by the tablespoon into a simmering vegetable soup. Cover and simmer (without peeking) for 10 minutes.

**6.** Remove the lid and serve.

# DEEP SEA SCALLOP SAUTÉ

PREPARATION: 10 MINUTES          COOKING: 15 MINUTES          YIELD: 4 SERVINGS

*Substitute the scallops with monkfish cut into 2-inch chunks. This dish is good over pasta twists or rice.*

2 tablespoons olive oil
2 cloves garlic, crushed or minced
1 red bell pepper, cut in 1-inch pieces
1 6-inch zucchini, cut julienne
6 scallions with green, sliced
1½ teaspoons grated lemon zest
½ teaspoon coarsely ground black pepper
1½ cups tomato sauce
¼ cup cream
1 pound deep sea scallops
1½ cups corn kernels
¼ cup chopped parsley

**1.** Heat the oil in a large skillet and sauté the garlic, red pepper, and zucchini for 3 minutes.

**2.** Add the scallions, grated lemon, and black pepper, and sauté 2 minutes.

**3.** Pour in the tomato sauce and cream and heat 3 minutes, until it begins to bubble gently.

**4.** Add the scallops and corn kernels, lower the heat if necessary, and simmer 5 minutes, until the scallops are cooked through.

**5.** Sprinkle each serving with chopped parsley.

# CROCKPOT CHICKEN THIGHS

PREPARATION: 10 MINUTES　　　COOKING: 4 HOURS ON HIGH　　　YIELD: 4 SERVINGS

*So easy, so good, and the best thing about it is that when you sit down to eat it, you have the wonderful feeling of having done nothing. I don't thicken the juices because I serve it over baked potato or rice. If you like thick gravy, stir 2 tablespoons cornstarch into a tablespoon of water and stir into the pot 5 minutes before you're ready to serve.*

3 stalks celery, sliced very thin
1 medium onion, sliced very thin
1 green pepper, cut in ½-inch pieces
8 chicken thighs, skin and fat removed
6 cloves garlic, crushed or minced
1 teaspoon dried thyme
¼ teaspoon ground allspice
½ teaspoon coarsely ground black pepper
1 cup tomato juice
1 cup apple or carrot juice, or chicken stock
2 cups corn kernels

**1.** Preheat the Crockpot.

**2.** Place all the vegetables on the bottom of the Crockpot and place the chicken thighs on top.

**3.** Sprinkle with the garlic, thyme, allspice, and black pepper.

**4.** Pour over the juices.

**5.** Cover the pot and cook 3 hours. Add the corn kernels and cook 1 hour longer.

# MEATBALLS AND SPAGHETTI SAUCE

PREPARATION: 10 MINUTES (20 IF NOT USING A PROCESSOR)　　　　YIELD: 4 SERVINGS
COOKING: 25 MINUTES

*I make both the meatball mixture and the sauce vegetables in the food processor. If you don't want to do that, just chop the vegetables very fine and throw the corn into the blender a little at a time. Serve with 1 pound of spaghetti, cooked according to package directions.*

1 tablespoon vegetable oil
2 small carrots, cut in 1-inch pieces
1 medium onion, cut in 8 pieces
1 large green bell pepper, cut in 8 pieces
4 cloves garlic
4 cups tomato sauce (I use whole canned tomatoes puréed)
1 tablespoon dried basil
1 teaspoon dried oregano
¼ teaspoon coarsely ground black pepper
4 large slices bread (I use Arnold's Branola Oatmeal)
4 cloves garlic
½ cup parsley, stems removed
1 cup corn kernels
1 teaspoon dried thyme
1 teaspoon dried leaf sage
½ teaspoon coarsely ground black pepper
½ pound ground beef or pork
1 large egg

**1.** Heat the oil in a large skillet over medium.

**2.** Process the carrots, onion, green pepper, and garlic with an on-off motion for 40 seconds. Add to the skillet and sauté for 2 minutes.

**3.** Stir in the tomato sauce, basil, oregano, and black pepper.

**4.** Simmer 5–10 minutes while you're preparing the meatballs.

**5.** Place the bread slices, garlic, parsley, corn, thyme, sage, and pepper in the food processor and process for 1 minute.

**6.** Add the meat and egg and process on-off for 40 seconds to combine all ingredients.

**7.** Form the meatballs (about 20) and sauté them in a large skillet for 2 minutes, turning them to brown each side.

**8.** Add to the tomato sauce and cook gently for 10–15 minutes.

When these hard-working folks relaxed at the end of the day, they sipped on corn juice, corn cider, corn beer, corn wine, or corn liquor. Simple corn whiskey stills were commonplace in the 1700s, and by the late eighteenth century, whiskey-making was so popular that United States revenue officers were sent around the country to enforce a federal tax on all distillers. Some small-time distillers hid out in the backwoods, making "moonshine" and "likker"; others fled out of state. A Baptist minister named Elijah Craig established his still in Bourbon County, Kentucky, and called his brew Bourbon County Whiskey. Thus, bourbon whiskey was born in 1789. Today, Kentucky remains the heart of the American whiskey industry and is the major source of bourbon.

# STUFFINGS AND RELISHES

## CHAPTER 8

*Never thrust your sickle into another's corn.*

—PUBLIUS SYRUS, MAXIM 593

**U**sually a combination of bread, vegetables, meats, and nuts, stuffings can be made from just about anything. For example, fruit stuffings made with dried prunes or apricots, or from fresh chopped apple, are particularly delicious for accompanying goose or duck. Stuffings made with nuts, sausage meat, and cornbread are more suited for turkey and chicken. Whole corn, though, can be added to many different kinds of stuffings without greatly changing the flavor. Don't add much more than a cup or the stuffing won't bind together properly.

You don't have to wait until you roast a whole bird. Stuffings make a hearty side dish for baked chicken, a sautéed slice of thick ham, sausages, pork roast, and venison.

Pennsylvania Dutch country is famous for its corn relish and its sweet-and-sour chow-chow pickles. The chow-chow, made from as many as fourteen vegetables, is so thick that it is often served as a side dish. A major crop in the Pennsylvania farmlands, it goes without saying that corn is one of the main ingredients. Besides flavor, corn adds color and texture (as long as you don't overcook the relish), and one great advantage is that it comes in small pieces. The most you may have to do is remove it from the cob.

Many relishes are cooked briefly, bottled, and processed in a hot water bath. If you are not putting up vast quantities and shelf-life is of no concern, then you can get away with making a quick refrigerator relish. It will taste just as good as the processed, but you'll have to eat it within two weeks.

# CHESTNUT AND CORN BREAD STUFFING

PREPARATION: 60 MINUTES     BAKING SIDE DISH: 40 MINUTES     YIELD: 12–16 SERVINGS

*Good as a side dish, or stuff into the crop and cavity of a 14-pound turkey and roast in a preheated 350° oven for 3½–4 hours, basting frequently.*

1 corn bread recipe (page 114)
½ pound chestnuts
4 tablespoons olive or vegetable oil
1 large onion, chopped fine
2 large ribs celery, chopped fine
¼ cup chopped parsley
1 tablespoon dried sage leaves
1 teaspoon dried thyme leaves
½ teaspoon coarsely ground black pepper
½–¾ cup chicken stock

**1.** Prepare the corn bread according to the recipe on page 114. Cool on a wire rack when baked.

**2.** While the corn bread is baking, cook the chestnuts. Cut an X on the flat side of each chestnut and place in a saucepan of cold water. Bring to a boil, reduce the heat immediately and simmer for 15 minutes. Drain and cool until they can be handled.

**3.** Preheat the oven to 350°F.

**4.** Remove the chestnut shells and peel off the brown skin with a sharp paring knife. Purée in a ricer, food mill, or food processor.

**5.** Heat the oil in a large skillet and sauté the onion and celery for 5 minutes.

**6.** Stir in the parsley, sage, thyme, pepper, and puréed chestnuts.

**7.** Crumble the corn bread and add to the mixture in the skillet.

**8.** Moisten with the chicken stock.

**9.** Bake for 40 minutes.

# WHOLE CORN STUFFING

PREPARATION: 10 MINUTES　　　　　BAKING: 30 MINUTES　　　　　YIELD: 6–8 SERVINGS

*This amount of stuffing will fill the cavities of two 4-pound chickens or one 8-pound goose or turkey, or you can serve it as a side dish.*

2 tablespoons olive or vegetable oil
1 medium onion, chopped fine
1 sweet red bell pepper, chopped fine
1 clove garlic, crushed
1 cup corn kernels
9 slices hearty whole wheat bread
1 teaspoon dried thyme leaves
½ teaspoon coarsely ground black pepper
¼ cup liquid (stock, apple or tomato juice)
1 large egg, beaten

**1.** Preheat the oven to 350°F.

**2.** Heat the oil in a skillet and add the onion, pepper, and garlic. Sauté for 5 minutes.

**3.** Add the corn and cook 2 minutes longer. Remove from the heat.

**4.** Make breadcrumbs with the whole wheat slices (about 3 cups) and add to the skillet with the thyme and pepper.

**5.** Stir in the liquid and beaten egg.

**6.** Bake for 30 minutes.

# PECAN CORN BREAD STUFFING

PREPARATION: 45 MINUTES     BAKING SIDE DISH: 40 MINUTES     YIELD: 12–16 SERVINGS

*Serve as a side dish, or stuff into the crop and cavity of a 10–12-pound turkey and roast in a preheated 350° oven for 3½ hours.*

1 corn bread recipe (page 114)
3 tablespoons olive or vegetable oil
6 large scallions, including the green,
   sliced thin
1 sweet red bell pepper, chopped fine
¼ cup chopped flat parsley
1 cup chopped pecans (or walnuts)
1 tablespoon dried basil
½ teaspoon coarsely ground black pepper
¼–⅓ cup apple juice
1 large egg, beaten

**1.** Prepare the corn bread according to the recipe on page 114. Cool on a wire rack.

**2.** Preheat the oven to 350°F.

**3.** Heat the oil in a large skillet and sauté the scallions and red pepper for 5 minutes.

**4.** Remove from the heat and stir in the parsley, pecans, basil, and black pepper.

**5.** Crumble the corn bread and add to the pecan mixture. Stir to combine.

**6.** Moisten with the apple juice and bind with the beaten egg.

**7.** Bake for 40 minutes.

# REFRIGERATED CORN RELISH

PREPARATION: 10–20 MINUTES                    YIELD: 6 CUPS

*This quick relish can remain refrigerated for up to two weeks.*

½ cup apple cider vinegar
¼ cup lemon juice
¾ cup sugar
½ teaspoon turmeric
½ teaspoon coarsely ground black pepper
½ teaspoon celery seed
1 large red bell pepper, diced
6 large scallions, including greens, sliced thin
4½ cups canned corn kernels (or 9 ears of corn, scraped)

**1.** Place the vinegar, lemon juice, sugar, and seasonings in a small saucepan. Bring to a boil and simmer 2–3 minutes.

**2.** Place the red pepper, scallions, and corn in a bowl and cover with the hot vinegar mixture.

**3.** Toss to combine and refrigerate for 1 day before serving.

# FRESH CORN RELISH

12 ears of corn
1 large red bell pepper, diced
1 large green bell pepper, diced
1 large onion, chopped fine
1½ cups apple cider vinegar
¼ cup water
1½ cups sugar
1 teaspoon dry mustard
1 teaspoon curry powder (optional)
1 teaspoon celery seeds
1 tablespoon sea salt

**1.** Using a sharp knife and working from top to bottom, remove the kernels from the cobs. (There will be approximately 6 cups of kernels and liquid.) Place in a large enamel or stainless steel kettle.

**2.** Add the rest of the ingredients and stir to combine.

**3.** Bring to the boiling point, lower the heat, and simmer for 15 minutes. Turn off the heat.

**4.** With the help of a wide-mouthed funnel, spoon the mixture into sterilized, hot canning jars, leaving a ½-inch space at the top.

**5.** Wipe the jar rims and cover with sterilized lids. Screw down the lids tightly.

**6.** Place the jars, not touching, in a canning kettle and cover completely with hot water. Bring to a boil and simmer for 10 minutes. Remove from the hot water bath with canning tongs and allow to cool.

**7.** Store in a cool, dark place for 3–4 weeks before using.

# BREADS, MUFFINS, AND GRIDDLE CAKES

CHAPTER 9

*Corne, which is the staffe of life . . .*
—EDWARD WINSLOW (1595–1655)
GOOD NEWS FROM NEW ENGLAND
(LONDON, 1624)

# YEAST AND QUICK BREADS

**A**ll manner of breads can be made from finely ground cornmeal, water and melted fat. There are ceremonial breads such as Hopi *chukuviki*, a corn bread wrapped in corn husks and steamed, and Hopi *piki* bread, two paper-thin pancakes rolled together. Corn pone—small, mounded, oven-baked or fried breads—originated from Algonquin *appome* or *apan*. The Narraganset Indians wrapped small mounds of stiff batter in leaves and baked them in hot ashes. These were called ash cakes, or hoe, by the settlers; stick cakes were baked on the end of a long hoe or stick; and jonny (or journey) cakes were baked beside an open fire. From the Indians the settlers also borrowed Indian or hasty pudding (cornmeal, milk, and spices sweetened with molasses), boiled and fried mushes, corn muffins, corn dodgers (small and extremely dense rounds of batter either baked or fried), and pudding-like spoon bread.

The reason there are so many different types of flat corn breads and cakes is because cornmeal and corn flour contain very little gluten. In other words, they have no leavening power and the result is a close texture with no elasticity.

Colonial women had their own tricks for making lighter breads. They made "starters" by mixing flour, water (sometimes from the boiled potatoes), and sugar. This was then left open to the air to attract wild yeast spores. The homemade "starter" was added to the bread flour, and a little was saved to start the next batch of bread. On other occasions they would add brewer's yeast, a by-product of their beer making.

Other methods of leavening included using beaten egg whites, sour milk, and pearl ash, an early form of baking powder. It was not until the 1860s that the Royal Company manufactured commercial baking powder and baked goods became light and airy concoctions.

The corn breads that are made exclusively with cornmeal are much flatter and denser, even with the addition of baking powder.

For more quick bread variations, use the muffin recipes (pages 124–133) and bake in pans instead of cups. Then, again, you can take the quick bread recipes in this chapter and turn them into muffins.

---

# ANADAMA MOLASSES YEAST ROLLS

PREPARATION: 30 MINUTES          BAKING: 20 MINUTES          YIELD: 18 ROLLS

*This is one of my favorite yeast bread recipes. It's hearty and flavorful and because of the no-knead batter method, it takes less than an hour from the mixing bowl to the table.*

2 packages active dry yeast (2 tablespoons)
⅓ cup non-fat dry milk
⅓ cup molasses
¼ cup vegetable oil
1 cup cornmeal
2 cups hand-hot water (120°F.–130°F.
  degrees)
1½ cups whole wheat flour
1½ cups all-purpose flour

**1.** Preheat the oven to 375°F. and grease 18 muffin cups.

**2.** Place the yeast, dry milk, molasses, oil, and cornmeal in a large mixing bowl.

**3.** Add the hot water and beat together for 1 minute.

**4.** Combine the whole wheat and all-purpose flours and beat 2 cups into the yeast mixture. Beat for 2 minutes with an electric mixer or 4 minutes by hand.

**5.** Stir in the remaining cup of flour and beat by hand for 2 minutes. The batter will be sticky.

**6.** Using a wet or oily spoon, fill the muffin cups with the batter.

**7.** Cover with a clean cloth and allow to rise for 20 minutes in a warm place.

**8.** Bake for 20 minutes, or until a skewer inserted into the center comes out clean. The rolls will sound hollow when rapped with the knuckles.

**9.** Remove from the cups and cool on a wire rack. Delicious when served warm.

# POTATO-HERB BATTER BREAD

PREPARATION: 30 MINUTES       BAKING: 40 MINUTES       YIELD: 2 LOAVES

*When I'm not in a hurry, I make this in a 2½-quart casserole for the effect of serving a large, round loaf. Done that way, it takes 1 hour in the oven. This bread slices nicely even when warm.*

2 packages active dry yeast
  (2 tablespoons)
2 tablespoons honey
¼ cup vegetable oil
1 teaspoon dried basil
1 teaspoon dried oregano
¾ cup dry mashed potatoes (don't use
  leftover mashed potatoes containing
  milk)
1 cup cornmeal
2 cups hand-hot water (use potato water
  if desired)
1 large egg, beaten (optional)
2½ cups all-purpose flour

**1.** Preheat the oven to 375°F. and grease two 7½" × 3½" × 2¼" pans.

**2.** Place the yeast, honey, oil, basil, oregano, mashed potatoes, and cornmeal in a large mixing bowl.

**3.** Add the hot water (120°F.–130°F. degrees) and beat together for 1 minute.

**4.** Stir in the egg and 1½ cups flour. Beat with an electric mixer for 2 minutes or by hand for 4 minutes.

**5.** Beat in the remaining 1 cup flour by hand for 2 minutes. If the batter feels stiff, beat in 1 extra tablespoon oil.

**6.** Using a wet or oil-coated spoon, distribute the batter between the two loaf pans, cover with a clean towel, and allow to rise for 20 minutes in a warm place.

**7.** Bake 40 minutes, or until a skewer inserted in the center comes out clean. Fully baked loaves will sound hollow when rapped with the knuckles.

**8.** Remove from pans and cool on a wire rack.

# CORN KERNEL YEAST BREAD

PREPARATION: 30 MINUTES          BAKING: 40 MINUTES          YIELD: 2 LOAVES

*I use this as my basic bread recipe. Sometimes I use molasses instead of the honey and substitute nuts and raisins for the corn kernels and scallions. It's also delicious with 2–4 tablespoons of chopped fresh herbs.*

2 packages active dry yeast (2 tablespoons)
⅓ cup non-fat dry milk
2 tablespoons honey
¼ cup vegetable oil
¼ cup thinly sliced scallion greens (or 2 tablespoons chopped chives)
1½ cups corn kernels
2 cups hand-hot milk (120°F.–130°F. degrees)
1½ cups all-purpose flour
1½ cups whole wheat flour

**1.** Preheat the oven to 375°F. and grease two 7½″ × 3½″ × 2¼″ loaf pans.

**2.** Place the yeast, dry milk, honey, oil, scallions, and corn kernels in a large mixing bowl.

**3.** Pour in the hot milk (it should not feel "scalding" to the touch, just hot) and beat together for 1 minute.

**4.** Combine the flours and beat 1½ cups into the yeast mixture. Use an electric mixer for 2 minutes or beat by hand for 4 minutes.

**5.** Stir in the remaining flour and beat by hand for 2 minutes. If the batter feels too stiff, add an extra tablespoon of oil.

**6.** Using a wet or oiled spoon, transfer the batter into the loaf pans, cover with a clean towel, and allow to rise in a warm place for 20 minutes.

**7.** Bake for 40 minutes, or until a skewer inserted in the center comes out clean. The loaves should sound hollow when rapped with the knuckles. Cool on a wire rack.

# WHOLE WHEAT CORN LOAF

PREPARATION: 2 HOURS                    BAKING: 35 MINUTES                    YIELD: 1 LOAF

*This is a crumbly bread but a satisfying stick-to-the-ribs one. You might want to jazz up the flavor by adding 1 cup chopped ham, ½ cup chopped onion, ½ cup Parmesan cheese, or 1 teaspoon cumin powder.*

2 tablespoons vegetable oil or melted
    shortening
2 tablespoons sugar
¼ cup non-fat dry milk
1¼ cups hand-hot water (120°F.–130°F.
    degrees)
1 cup cornmeal
1½ cups all-purpose flour
1½ packages active dry yeast
1 cup whole wheat flour

**1.** Preheat oven to 375°F., grease a baking sheet, and sprinkle with cornmeal.

**2.** Combine the oil, sugar, dry milk, and hot water in a large mixing bowl.

**3.** Stir in the 1 cup cornmeal and ½ cup of all-purpose flour. Beat for 1 minute with an electric mixer or 2 minutes by hand.

**4.** Add the rest of the all-purpose flour and the yeast. Beat for 2 minutes with the electric mixer or 4 minutes by hand.

**5.** Mix in the whole wheat flour until the dough forms a soft ball. Cover and let rest for 5 minutes.

**6.** Turn onto a lightly floured surface and knead for 10 minutes—until the dough is smooth and elastic.

**7.** Place the dough in a greased bowl, cover, and allow to rise in a warm, draft-free area until it doubles in bulk in approximately 60 minutes.

**8.** Punch down and shape into a round and place on the baking sheet. Cover and let rise again for 30–45 minutes.

**9.** Bake in a preheated oven for 35 minutes, or until a skewer inserted in the center comes out clean. The loaf will sound hollow when tapped with the fingers.

# REGULAR CORN BREAD

PREPARATION: 10 MINUTES      BAKING: 20 MINUTES      YIELD: 12 SERVINGS

*I don't add salt to any of my recipes. If you want to perk this plain corn bread up a bit, add ½ teaspoon salt or 4 slices of crispy, crumbled bacon. I sometimes add 1 cup corn kernels, or ½ cup grated cheddar cheese plus ¼ cup sliced scallions.*

*Use this recipe for the stuffings on pages 102 and 104.*

1 cup all-purpose flour
1 cup yellow cornmeal
1 tablespoon baking powder
1–2 tablespoons sugar
1 cup low-fat milk
2 large eggs, beaten
¼ cup vegetable oil or melted shortening

**1.** Preheat the oven to 425°F. and grease an 8″ or 9″ × 2″ pan.

**2.** Combine the flour, cornmeal, baking powder, and sugar in a large mixing bowl.

**3.** Make a well in the center and pour in the milk, beaten eggs, and oil.

**4.** Stir until the dry ingredients are moistened and then spoon into the pan.

**5.** Bake 20 minutes, or until a skewer inserted in the center comes out clean.

# MEXICAN CORN BREAD

PREPARATION: 15 MINUTES          BAKING: 30 MINUTES          YIELD: 12 SLICES

*Serve this with a bowl of chili and you'll get rave reviews. For an extra spicy version, use Jack cheese with jalapeño peppers instead of the cheddar cheese.*

1¼ cups all-purpose flour
¾ cup cornmeal
2 teaspoons baking powder
1 teaspoon baking soda
1–2 tablespoons sugar
¼ cup low-fat milk
½ cup sour cream
¼ cup vegetable oil
1 large egg, beaten
½ cup chopped mild green chili peppers
1 cup grated cheddar cheese (4 ounces)
1 cup corn kernels

**1.** Preheat the oven to 400°F. and grease an 8- or 9-inch square baking pan.

**2.** Mix the flour, cornmeal, baking powder, baking soda, and sugar in a large mixing bowl.

**3.** Combine the milk, sour cream, oil, beaten egg, chilis, cheese, and corn.

**4.** Stir into the dry ingredients until barely moistened.

**5.** Spoon into the prepared pan and bake for 30 minutes, or until a skewer inserted in the center comes out clean. Serve warm.

# PUMPKIN NUT BREAD

PREPARATION: 10 MINUTES          BAKING: 50 MINUTES          YIELD: 12–15 SLICES

*If you don't have cornmeal, substitute whole wheat flour. This is a delicious tea bread.*

1 cup all-purpose flour
½ cup whole wheat flour
¾ cup cornmeal
2 teaspoons baking powder
1 teaspoon baking soda
2 teaspoons ground cinnamon
¼ teaspoon ground cloves
½ teaspoon ground nutmeg
⅓ cup brown sugar
½ cup honey
¼ cup vegetable oil or melted shortening
½ cup apple juice
2 large eggs
1 cup fresh or canned pumpkin purée
¾ cup chopped walnuts or pecans

**1.** Preheat the oven to 350°F. and grease a 9″ × 5″ × 3″ loaf pan.

**2.** Mix the flours, cornmeal, baking powder, baking soda, spices, and brown sugar in a large mixing bowl.

**3.** Beat together the honey, oil, apple juice, eggs, and pumpkin purée.

**4.** Stir gently into the dry ingredients until barely moistened.

**5.** Fold in the nuts and spoon into the prepared pan.

**6.** Bake for 50 minutes, or until a skewer inserted in the center comes out clean.

**7.** Cool in the pan for 10 minutes. Remove and cool on a wire rack.

# BOSTON BROWN BREAD

PREPARATION: 10 MINUTES          BAKING: 40 MINUTES          YIELD: 2 LOAVES

*This is an American classic, although older recipes call for rye flour and require steaming the batter.*

1½ cups all-purpose flour
1 cup whole wheat flour
1 cup cornmeal
2 teaspoons baking powder
1 teaspoon baking soda
2 large eggs
2 cups buttermilk
1 cup molasses
1½ cups dark raisins

**1.** Preheat the oven to 350°F. and grease two 1-pound coffee cans or two 7½″ × 3½″ × 2¼″ loaf pans.

**2.** Mix the flours, cornmeal, baking powder, and baking soda in a large bowl.

**3.** Beat together the eggs, buttermilk, and molasses. Stir in the raisins.

**4.** Combine with the dry ingredients and blend together.

**5.** Spoon into the greased pans and bake for 40 minutes, or until a skewer inserted in the center comes out clean.

**6.** Cool 10 minutes in the pan. Remove and cool on a wire rack.

# FOUR-GRAIN CORN BREAD

PREPARATION: 10 MINUTES          BAKING: 20 MINUTES          YIELD: 12 SLICES

*This is one of my favorite recipes for bread or muffins. It's got a crunchy texture and a flavor that makes it ideal for eating on its own (when it's warm, you don't even need butter), at break-fast time, or to accompany a savory meal at lunch or dinner.*

2¾ cups cornmeal
¾ cup all-purpose flour
½ cup rolled oats
½ cup grape nuts
1 tablespoon baking powder
½ cup chopped walnuts or pecans
2 large eggs
½ cup low-fat milk
6 tablespoons vegetable oil or melted
   shortening
¼ cup honey, molasses, or maple syrup

**1.** Preheat the oven to 400°F. and grease a 9″ × 2″ square or round pan.

**2.** Mix the cornmeal, flour, oats, grape nuts, baking powder, and walnuts in a large bowl.

**3.** Beat together the eggs, milk, oil, and honey.

**4.** Combine with the dry ingredients until barely moistened.

**5.** Spoon into the greased pan and bake 20 minutes, or until a skewer inserted in the center comes out clean.

**6.** Leave in the pan 5 minutes. Remove from the pan and cool on a wire rack or eat warm.

# BERRY BREAD

PREPARATION: 15 MINUTES          BAKING: 25 MINUTES          YIELD: 12 SLICES OR CUPCAKES

*This is a cake-like bread and is guaranteed to disappear very fast. Serve warm or cool with ice cream or whipped cream. Or turn it into a cupcake recipe and frost with cream cheese and confectioners sugar.*

1 cup all-purpose flour
1 cup cornmeal
⅓ cup sugar
1 tablespoon baking powder
½ teaspoon baking soda
½ cup melted butter blend or margarine
½ cup berry yogurt
1 large egg
1 cup berries or sliced strawberries

**1.** Preheat the oven to 400°F. and grease a 9″ × 2″ square or round baking pan.

**2.** Mix the flour, cornmeal, sugar, baking powder, and baking soda in a large bowl.

**3.** Beat together the melted shortening, yogurt, and egg.

**4.** Stir into the dry ingredients until just combined.

**5.** Gently fold in the berries.

**6.** Spoon into the greased pan and bake for 25 minutes, or until a skewer inserted in the center comes out clean.

**7.** Leave in the pan 5 minutes. Remove and cool on a wire rack.

# COTTAGE CHEESE BISCUITS

PREPARATION: 15 MINUTES          BAKING: 15 MINUTES          YIELD: 8 WEDGES

*These are something like British scones but a bit more crumbly. They are best when served warm or no more than one day old. Increase or decrease the sugar depending on whether you want to eat them as a bread with dinner or with preserves for breakfast. For a savory biscuit, add 1 tablespoon mixed herbs.*

2 cups all-purpose flour
1 cup cornmeal
1 tablespoon baking powder
½ teaspoon baking soda
2–4 tablespoons sugar
¼ cup butter blend or margarine
2 cups low-fat cottage cheese, blended
   smooth
2 eggs, beaten

**1.** Preheat the oven to 425°F. and grease a baking sheet.

**2.** Mix the flour, cornmeal, baking powder, baking soda, and sugar in a large bowl.

**3.** Cut the shortening into tablespoons and add to the flour mixture. Using a pasty blender or two knives, work it into the flour until coarse crumbs form.

**4.** Use a fork to stir in the cottage cheese and beaten eggs and mix until a ball forms.

**5.** Turn onto a floured surface and knead gently for about 3 minutes, or until smooth.

**6.** Pat the dough into a circle and flatten until it is about 1 inch thick. Cut into quarters and then halve the triangles so that there are 8 wedges.

**7.** Sprinkle the greased baking sheet with cornmeal and place the dough wedges in a loose circle.

**8.** Bake for 15 minutes, or until a skewer inserted in the center comes out clean.

# SAVORY BISCUITS

PREPARATION: 15 MINUTES          BAKING: 15 MINUTES          YIELD: 6 WEDGES

*You decide what goes in these. Add 1 cup of whatever you have on hand: chopped ham, cooked sausage, corn kernels, or cooked chopped vegetables. I sometimes make these as tea scones: Increase the sugar to 2 tablespoons and eliminate the thyme and sage. Serve them with butter and jam.*

1½ cups all-purpose flour
½ cup cornmeal
1 tablespoon baking powder
1 tablespoon sugar
1 teaspoon dried thyme
1 teaspoon dried sage leaves, crumbled
¼ cup butter blend or margarine
¾ cup low-fat milk

**1.** Preheat the oven to 425°F. and grease a baking sheet.

**2.** Mix the flour, cornmeal, baking powder, sugar, thyme, and sage in a large bowl.

**3.** Cut the shortening into tablespoons and add to the flour mixture. Using a pastry blender or two knives, work in until the mixture resembles coarse crumbs. (If using ham, sausage, etc., combine at this point.)

**4.** Pour in the milk and stir with a fork until the dough forms a ball.

**5.** Turn onto a floured surface and knead gently for about 3 minutes until the dough is smooth.

**6.** Form into a ball and flatten into a circle 1 inch thick. Cut into 6 wedges.

**7.** Sprinkle the baking sheet with cornmeal and place the wedges in a loose circle. Bake 15 minutes, until a skewer inserted in the center comes out clean. Serve warm or very fresh.

# CHEESE CRACKERS

PREPARATION: 15 MINUTES      BAKING: 15 MINUTES      YIELD: 24 CRACKERS
(OR THEREABOUTS)

*If you've never thought about making your own crackers, now's the time with this simple recipe. Omit the cheese if desired.*

¾ cup all-purpose flour
1 cup cornmeal
¼ teaspoon baking soda
½ teaspoon coarsely ground black pepper
¾ cup grated cheddar or Monterey Jack cheese (try the jalapeño)
¼ cup vegetable oil
¼ cup low-fat milk or water

**1.** Preheat the oven to 375°F. Lightly grease two or three baking sheets and dust with cornmeal.

**2.** Combine the flour, cornmeal, baking soda, black pepper, and grated cheese in a large mixing bowl.

**3.** Stir in the oil and milk with a fork until the dough forms a ball, adding a drop more liquid if necessary.

**4.** Turn onto a floured surface and knead for about 3 minutes.

**5.** Divide the dough in half and roll out each half until it's between ⅟₁₆ and ⅛-inch thick.

**6.** Cut into 1½-inch squares and place on the baking sheets.

**7.** Bake 15 minutes. Remove from the oven and cool on the baking sheets.

# MUFFINS

**A**s much as I love hearty, whole-grain yeast breads, muffins run a close second. And because they are so easy and fast to put together, I eat these miniature breads more than any other. Also, the variations on a basic recipe are practically endless. You can pack a lot of nutrition into muffins by adding half a cup of nuts, fruit, chopped vegetables, cheese, fruit juices, or anything else you and your family enjoy. To vary the flavor, try adding 1–2 tablespoons grated orange rind and substituting orange juice for the milk. Add 1 cup mashed ripe banana, or put a peanut butter surprise in the center of each muffin. It's hard not to go on and on about muffin possibilities. I'll end by saying: Experiment, experiment!

One of the great things about muffins is that you can eat fresh "bread" that only takes about 30 minutes from start to finish. The best muffins are warm muffins.

Any muffin recipe, of course, can be turned into a quick bread by simply using a square 8-inch or 9-by-2-inch pan.

When mixing muffins, the dry ingredients should be barely moistened. A lumpy batter results in tender muffins.

I frequently bake muffins in a microwave oven. To do this, I use either a special 6-cup microwave muffin pan or six 6-ounce custard cups filled ¾ full (and then repeat the procedure with the remaining batter) and cook uncovered on the HIGH setting for 3 minutes. If you don't have a rotating plate, cook on HIGH for 2 minutes, rotate the plate a half-turn and cook on HIGH for 1 minute longer. Microwave muffins will continue to cook after they have been removed from the oven. They have more of a pudding texture than regular muffins but are, nevertheless, quite delicious. Shaving 15 minutes off a recipe can sometimes mean the difference between breakfast or no breakfast.

# CORNMEAL MUFFINS

PREPARATION: 10 MINUTES          BAKING: 20 MINUTES          YIELD: 12 MUFFINS

*These are your basic cornmeal muffins—low on sugar and shortening and containing absolutely no vanilla flavoring. In other words, they make a superb, plain muffin to accompany a savory meal. For a richer and less crumbly muffin, increase the ingredients to: 2 large eggs, 6 tablespoons shortening, and 4 tablespoons sugar. For Cheese Cornmeal Muffins, simply add ½–¾ cup grated cheddar cheese in step 3. Mexican Cornmeal Muffins can be made by adding ½ cup chopped green chili peppers (mild or hot jalapeño) in step 3.*

1 cup yellow cornmeal
1 cup all-purpose flour
1 tablespoon baking powder
3 tablespoons sugar
1 large egg
1 cup low-fat milk
¼ cup melted margarine or vegetable oil

**1.** Preheat the oven to 425°F. and grease a 12-cup muffin pan or line with paper cups.

**2.** Mix the cornmeal, flour, baking powder, and sugar in a medium mixing bowl.

**3.** Beat together the egg, milk, and shortening. Make a well in the center of the flour mixture and pour in the liquids.

**4.** Stir together until the ingredients are barely combined. Spoon into the muffin cups and bake for 20 minutes.

# DOUBLE CORN MUFFINS

PREPARATION: 10 MINUTES          BAKING: 20 MINUTES          YIELD: 12 LARGE MUFFINS

1 cup cornmeal
1 cup all-purpose flour
1 tablespoon baking powder
½ teaspoon baking soda
¼ teaspoon ground cinnamon
¼ teaspoon ground nutmeg
¼ cup honey
½ cup butter blend or margarine, melted
2 large eggs
1 cup yogurt
1 cup corn kernels, fresh, frozen, or canned
  (drained of all liquid)
½ cup chopped walnuts (optional)

**1.** Preheat the oven to 425°F. and grease a 12-cup muffin pan or line with paper cups.

**2.** Mix the cornmeal, flour, baking powder, baking soda, cinnamon, and nutmeg in a large bowl. Make a well in the center.

**3.** Place the honey, melted shortening, eggs, yogurt, and corn kernels in a blender and process on "grate" for 1 minute. Pour into the center of the dry ingredients and stir until barely combined.

**4.** Stir in the chopped walnuts gently. Leave the mixture lumpy.

**5.** Spoon into the prepared muffin cups, almost filling each cup, and bake 20 minutes, or until a skewer inserted in the center comes out clean.

# BLUEBERRY MUFFINS

PREPARATION: 10 MINUTES      BAKING: 20 MINUTES      YIELD: 12 LARGE MUFFINS

*Don't hesitate to use other fresh or frozen fruit in this recipe. Instead of blueberries, substitute 1 cup raspberries, chopped apple, cranberries, plums, or peaches. When using a tart fruit, you will probably opt for ½ cup sugar.*

1 cup all-purpose flour
1 cup cornmeal
1 tablespoon baking powder
½ teaspoon baking soda
⅓–½ cup sugar
⅓ cup vegetable oil, melted butter blend, or margarine
¾ cup low-fat milk
2 large eggs
1 cup blueberries

**1.** Preheat the oven to 425°F. and grease a 12-cup muffin pan or line with paper cups.

**2.** Mix the flour, cornmeal, baking powder, baking soda, and sugar in a large bowl. Make a well in the center.

**3.** Beat together the shortening, milk, and eggs. Pour into the center of the dry ingredients and stir until barely combined.

**4.** Fold in the blueberries without overmixing the batter.

**5.** Spoon into the muffin cups and bake for 20 minutes, or until a skewer inserted in the center comes out clean.

# BLACKBERRY WHOLE WHEAT MUFFINS

PREPARATION: 10 MINUTES          BAKING: 25 MINUTES          YIELD: 12 LARGE MUFFINS

*If you can beat the birds to the blackberry patch, all you'll need is 1 cup of the ripe fruits. If you gather more, freeze them for future baking days.*

½ cup cornmeal
1 cup whole wheat flour
½ cup all-purpose flour
1 tablespoon baking powder
½ teaspoon baking soda
⅓ cup vegetable oil
⅓–½ cup honey
¾ cup apple juice
2 large eggs
1 cup blackberries

**1.** Preheat the oven to 425°F. and grease a 12-cup muffin pan or line with paper cups.

**2.** Mix the cornmeal, flours, baking powder, and baking soda in a large bowl. Make a well in the center.

**3.** Beat together the vegetable oil, honey, apple juice, and eggs. Pour into the center of the dry ingredients and stir until barely combined.

**4.** Gently fold in the blackberries so that the batter retains a lumpy texture.

**5.** Spoon into the muffin cups and bake for 25 minutes, or until a skewer inserted in the center comes out clean.

# MOLASSES APPLE MUFFINS

PREPARATION: 15 MINUTES          BAKING: 20 MINUTES          YIELD: 12 LARGE MUFFINS

½ cup cornmeal
½ cup whole wheat flour
1 cup all-purpose flour
1 tablespoon baking powder
½ teaspoon baking soda
1 teaspoon ground ginger
2 large eggs
½ cup apple juice
½ cup molasses
¼ cup vegetable oil
1 unpeeled Granny Smith apple, grated

**1.** Preheat the oven to 425°F. and grease a 12-cup muffin pan or line with paper cups.

**2.** Mix the cornmeal, flours, baking powder, baking soda, and ginger in a large bowl. Make a well in the center.

**3.** Beat together the eggs, apple juice, molasses, and vegetable oil. Stir in the grated apple.

**4.** Pour the liquids into the center of the dry ingredients and combine until the mixture is lumpy.

**5.** Spoon into the muffin cups and bake for 15–20 minutes, or until a skewer inserted in the center comes out clean.

# STRAWBERRY MUFFINS

PREPARATION: 10 MINUTES          BAKING: 20 MINUTES          YIELD: 12 MUFFINS

*If you don't have fresh strawberries, use strawberry preserves or anything else that would provide a delicious surprise bite.*

1 cup white cornmeal
1 cup all-purpose flour
1 tablespoon baking powder
½ teaspoon baking soda (if using buttermilk)
¼ cup sugar
⅓ cup vegetable oil or melted shortening
2 large eggs
¾ cup buttermilk or low-fat milk
⅓ cup mashed, fresh strawberries or strawberry preserves

**1.** Preheat the oven to 425°F. and grease a 12-cup muffin pan or line with paper cups.

**2.** Mix the cornmeal, flour, baking powder, baking soda, and sugar in a large bowl. Make a well in the center.

**3.** Beat together the vegetable oil, eggs, and buttermilk or low-fat milk. Stir into the dry ingredients until barely combined.

**4.** Half fill the muffin cups with the butter, then add 1 teaspoon of the mashed strawberries in the center. Cover with the remaining batter.

**5.** Bake for 20 minutes.

# SPICY WHOLE-GRAIN MUFFINS

PREPARATION: 10 MINUTES          BAKING: 20 MINUTES          YIELD: 12 MUFFINS

¼ cup bran flake cereal
¼ cup rolled oats
½ cup cornmeal
1 cup all-purpose flour or whole wheat
   flour
1 tablespoon baking powder
1½ teaspoons ground cinnamon
½ teaspoon ground ginger
½ cup raisins
2 large eggs
⅓ cup vegetable oil
⅓ cup molasses or honey
¾ cup low-fat milk

**1.** Preheat the oven to 425°F. and grease a 12-cup muffin pan or line with paper cups.

**2.** Mix the bran flakes, rolled oats, cornmeal, flour, baking powder, cinnamon, ginger, and raisins in a large bowl.

**3.** Beat together the eggs, oil, molasses, and milk. Combine with the dry ingredients until barely moistened and the batter is lumpy.

**4.** Spoon into the muffin cups and bake for 15–20 minutes, or until a skewer inserted in the center comes out clean.

# CARROT NUT MUFFINS

*If you like, you can substitute zucchini or apple for the carrots. (Add ½ cup grated cheddar cheese in step 3 to go with the apple.)*

1 cup all-purpose flour
½ cup whole wheat flour
½ cup cornmeal
⅓ cup brown sugar
1 tablespoon baking powder
½ teaspoon ground nutmeg
½ teaspoon ground cinnamon
2 large eggs
¾ cup low-fat milk
⅓ cup vegetable oil
1 cup grated carrot
½ cup chopped walnuts or pecans

**1.** Preheat the oven to 425°F. and grease a 12-cup muffin pan or line with paper cups.

**2.** Mix the flours, cornmeal, sugar, baking powder, and spices in a large bowl.

**3.** Beat together the eggs, milk, and oil. Stir in the grated carrot.

**4.** Pour into the dry ingredients and combine until barely moistened.

**5.** Fold in the chopped nuts and spoon into the muffin cups.

**6.** Bake for 20 minutes, or until a skewer inserted in the center comes out clean.

# BACON SCALLION MUFFINS

PREPARATION: 12 MINUTES          BAKING: 20 MINUTES          YIELD: 12 LARGE MUFFINS

4 slices bacon
1¼ cups all-purpose flour
¾ cup yellow cornmeal
1 tablespoon baking powder
1 cup low-fat milk
1 large egg
¼ cup vegetable oil
½ cup finely sliced scallions, or ¼ cup
  chopped chives
1 cup corn kernels (2 ears corn)

**1.** Preheat the oven to 425°F. and grease a 12-cup muffin pan or line with paper cups.

**2.** Cook the bacon until brown. Drain on paper towels.

**3.** Mix the flour, cornmeal, and baking powder in a large bowl.

**4.** Beat together the milk, egg, and oil. Stir in the sliced scallions and corn kernels. Pour into the dry ingredients and stir until barely moistened and the batter is lumpy.

**5.** Spoon into the muffin cups and bake for 20 minutes, or until a skewer inserted in the center comes out clean.

# STEVE COLEMAN'S VEGETARIAN MUFFINS

PREPARATION: 10 MINUTES          BAKING: 20 MINUTES          YIELD: 8 (TO 12) MUFFINS

*Steve Coleman is president of my Bedford Audubon Society chapter and director-naturalist at a local Nature Conservancy sanctuary. I participated in a naturalist training program with him last year which he made doubly exciting by baking giant health muffins before each session. There were Bran and Molasses, Whole Wheat and Wild Blackberry, and, of course, Corn.*

*Following is Steve's basic recipe. He uses 2 cups of whole wheat flour and makes variation changes with the 1 cup measurement using unprocessed bran, rolled oats, wheat germ, or cornmeal. He also likes to add nuts, raisins, and fresh fruit. As a pretty strict vegetarian, Steve does not use animal dairy products.*

2 cups whole wheat flour (or 1 cup all-purpose plus 1 cup whole wheat)
1 cup yellow cornmeal
2 teaspoons baking powder
1 teaspoon baking soda
½ cup vegetable oil
½ cup liquid sweetener (honey, barley malt, rice syrup, or molasses)
1¾ cups fruit juice, soy milk, or water (when using all-purpose flour or unprocessed bran, wheat germ, etc., try 1½ cups liquid)

**1.** Preheat the oven to 425°F. and grease 8 or 12 muffin cups depending on whether you want regular or huge muffins. (To get big muffins, fill the cups absolutely full with the batter or use 6-ounce custard cups.)

**2.** Mix the dry ingredients in a large bowl.

**3.** Combine the liquids and stir into the dry ingredients until just moistened.

**4.** Spoon the lumpy batter into the muffin cups and bake for 20 minutes, or until a skewer inserted in the center comes out clean.

# GRIDDLE CAKES

**D**epending on your ancestry and place of birth, your family may eat dodgers, pones, hush puppies, oysters, toads, fritters, pancakes, flat cakes, or johnny cakes. Some of these names were derived from the shape or consistency of the product. Dodgers, for example, are thought to be so named from their denseness; if used as flying missiles, they could wreak much damage. Corn oysters were said to cook in the shape of oysters; and hush puppies evolved from a corn and hash mixture or some say they were thrown down to hush the puppies. Johnny cakes originated in New England and have gained a reputation for their simplicity and flatness. Their shape made them ideal for taking along on a journey, so these flat cakes became known as journey cakes, and finally johnny or jonny cakes. Whether you cook fritters, pancakes, or oysters, they are all delicious if you're crazy about corn.

# CORN OYSTERS

PREPARATION: 10 MINUTES    COOKING: 3 MINUTES PER BATCH    YIELD: 4 SERVINGS

*There's practically nothing to these airy little cakes. Eat them for breakfast with maple syrup or serve them as a vegetable side dish.*

2 large eggs, separated
1½ cups corn kernels, crushed (if using
   fresh corn, grate it off the cob)
1 tablespoon low-fat milk
¼ cup all-purpose flour
¼ teaspoon coarsely ground pepper (OR
   1 tablespoon sugar)
2 tablespoons vegetable oil

**I.** Beat the egg yolks in a large bowl and stir in the corn and milk.

**2.** Add the flour and pepper (or sugar) and combine.

**3.** Beat the egg whites until stiff and fold them into the corn batter.

**4.** Heat the oil in a large skillet. When hot, drop in tablespoons of the batter (don't crowd the pan) and cook approximately 1½ minutes per side, until golden crisp and puffy.

**5.** Add a little more oil to the pan if necessary for each batch. Serve hot.

# REAL CORN OYSTERS

*Not only do these look like oysters, they also taste like oysters!*

¾ cup cornmeal
¾ cup all-purpose flour
2 teaspoons baking powder
½ teaspoon coarsely ground black pepper
1¼ cups low-fat milk
2 large eggs, separated
1 cup chopped oysters
¼ cup vegetable oil

**1.** Place the cornmeal, flour, baking powder, pepper, milk, and egg yolks in a large mixing bowl or a blender and blend together.

**2.** Stir in the oysters.

**3.** Beat the egg whites in a large bowl until stiff. Fold into the corn mixture.

**4.** Heat the oil in a large skillet and drop in about 2 tablespoons of the batter for each corn oyster. Don't overcrowd the pan.

**5.** Fry about 1½–2 minutes, on each side, until the oysters are golden brown. Serve hot.

# BUTTERMILK CORN PANCAKES

PREPARATION: 5 MINUTES     COOKING: 5 MINUTES PER BATCH     YIELD: 4 SERVINGS

*If you find these too corny, try using half all-purpose flour and half cornmeal. You can also use blue cornmeal and substitute chopped apple for the corn.*

1¼ cups buttermilk
1 cup cornmeal
1 large egg
½ teaspoon baking soda
1 cup corn kernels
2 tablespoons vegetable oil

**1.** Place the first four ingredients in a blender or mixing bowl and blend together.

**2.** Stir in the corn.

**3.** Heat 1 tablespoon of the oil in a large skillet and pour in about 2 tablespoons of batter for each pancake.

**4.** Cook over medium heat for 1–2 minutes a side, or until golden.

**5.** Add the second tablespoon of oil to the skillet, if desired, for the remaining batches. Serve hot.

# BASIC PANCAKES OR WAFFLES

PREPARATION: 2 MINUTES COOKING: 5 MINUTES PER BATCH YIELD: 4 SERVINGS

*This is my regular recipe, and I use whatever cornmeal I have on hand, which may be white, yellow, or blue—preferably stoneground. This is also my waffle batter. Sometimes I add 1 cup creamed corn if I'm making pancakes.*

*I like to serve stuffed pancakes occasionally as the main meal. I make the pancakes with ¼ cup batter and use them in place of the tortillas for Chicken Enchiladas (page 78) or instead of the patty shells for Curried Chicken and Corn (page 93). You can also add 1 cup diced ham and scallions to the recipe for Creamed Corn (page 54) and use that as a filling.*

1⅓ cups low-fat milk
½ cup cornmeal
¾ cup all-purpose flour
2 large eggs
2 tablespoons melted shortening or vegetable oil
¼ teaspoon ground nutmeg (optional)
2 tablespoons vegetable oil

**1.** Place all ingredients except the 2 tablespoons oil, in a blender and blend 1 minute, or beat 2 minutes by hand.

**2.** Heat 1 tablespoon of the oil in a large skillet and pour in about 2 tablespoons of batter for each pancake.

**3.** Cook over medium heat for 1–2 minutes each side, or until golden brown.

**4.** Add the second tablespoon of oil to the skillet, if desired, for the remaining batches of pancakes. (If making waffles, get the iron hot and cook about 2 minutes.)

# SOUR CREAM CORN PANCAKES

PREPARATION: 2 MINUTES      COOKING: 3 MINUTES PER BATCH      YIELD: 4 SERVINGS

*Serve a platter of these for brunch (with sausage on the side), and you'll have some very satisfied customers.*

¼ cup sour cream
2 large eggs
2 tablespoons vegetable oil
½ cup all-purpose flour
1 cup corn kernels, crushed, or, if using
　　fresh corn, grated off the cob
2 tablespoons vegetable oil

**1.** Beat the first five ingredients together.

**2.** Heat 2 tablespoons oil in a large skillet and drop the batter in 2 tablespoons at a time.

**3.** Cook over medium heat for 1½–2 minutes a side, until golden brown. Serve hot.

# VEGETABLE PANCAKES

PREPARATION: 5–10 MINUTES COOKING: 4 MINUTES PER BATCH      YIELD: 6 SERVINGS

*This is a great way to get children to eat vegetables. However, you can substitute 1 cup of fruit for the vegetables.*

½ cup all-purpose flour
½ cup cornmeal
1 tablespoon vegetable oil
1 large egg, beaten
1 cup low-fat milk
1 cup cooked, chopped vegetables; raw,
　　grated vegetables; or chopped fruit
2 tablespoons vegetable oil

**1.** Combine the first six ingredients in a mixing bowl.

**2.** Heat 1 tablespoon of the oil in a large skillet and drop in large spoonfuls of the batter.

**3.** Cook over medium heat for about 2 minutes each side, until golden brown.

**4.** Repeat with the remaining oil and batter. Serve hot.

# RHODE ISLAND JONNY CAKES

PREPARATION: 1 MINUTE      COOKING: 4 MINUTES PER BATCH      YIELD: 4 SERVINGS

*Eat jonny cakes with butter and maple syrup or molasses, ham, bacon, or sausage. In Rhode Island, they are also served with applesauce, creamed codfish, or chipped beef. There are two versions: Newport County and South County, both made from stoneground Rhode Island white-cap flint corn. Both claim to be the very best. Judge for yourself.*

## Newport County Jonny Cakes

PREPARATION: 1 MINUTE      COOKING: 4 MINUTES PER BATCH      YIELD: 4 SERVINGS

1 cup stoneground white cornmeal
½ teaspoon salt
1¾ cups cold milk
2 tablespoons vegetable oil or shortening

**1.** Mix the cornmeal, salt, and cold milk. The batter will be soupy.

**2.** Heat the oil on a griddle and ladle large spoonfuls of batter on top. The cakes should be about 5 inches in diameter.

**3.** Cook over medium heat for 2 minutes a side, until golden brown.

## South County Jonny Cakes

PREPARATION: 2 MINUTES      COOKING: 10 MINUTES PER BATCH      YIELD: 4 SERVINGS

1 cup stoneground white cornmeal
½ teaspoon salt
1 cup boiling water
1 teaspoon sugar or molasses (optional)
2 tablespoons vegetable oil or shortening

**1.** Combine the first four ingredients.

**2.** Heat the oil on a griddle and drop large spoonfuls of batter on top. The cakes should be about 3 inches in diameter.

**3.** Cook over medium heat for 6 minutes the first side and 4 minutes the second.

# ROSS EDWARDS' BLUE BLAZES HUSH PUPPIES

PREPARATION: 5 MINUTES          COOKING: 3 MINUTES A BATCH          YIELD: 35 HUSH PUPPIES

*This recipe is from Ross Edwards' Blue Corn Connection store. He recommends serving the hush puppies with fried fish, or as an appetizer. You can write to him at Blue Corn Connection, 8812 4th Street NW, Albuquerque, New Mexico 87114 for a mail order of the blue cornmeal.*

1 large egg
½ cup buttermilk
1¼ cups blue cornmeal
2 teaspoons baking powder
½ teaspoon salt
1 large clove garlic, crushed or minced
1 small onion, chopped fine
½ cup finely chopped green chilis, fresh,
　　frozen, or canned
¼–½ cup vegetable oil

**1.** In a medium mixing bowl, beat together the egg and buttermilk.

**2.** Stir in the cornmeal, baking powder, and salt.

**3.** Mix in the garlic, onion, and chilis. If using thawed or canned chilis, drain very well; if batter is too wet, add an extra tablespoon of cornmeal.

**4.** Heat the oil in a skillet (it should be about 1 inch deep) and drop the batter in by the teaspoon.

**5.** Cook about 1 minute on each side, until crisp and brown.

# CORN DODGERS

PREPARATION: 5 MINUTES COOKING TIME: 4 MINUTES PER BATCH          YIELD: 6 SERVINGS

*These can be dropped onto a greased baking sheet and baked in a 400°F. oven for 15 minutes. I prefer to fry them, however.*

2 cups cornmeal
1 teaspoon baking soda
¼ cup sugar
2 large eggs, beaten
1 cup plain yogurt, or 1 cup milk plus 1
    teaspoon vinegar
2 tablespoons vegetable oil or melted
    shortening
1 teaspoon vegetable oil

**1.** Sift the cornmeal and baking soda into a large bowl and add the sugar.

**2.** Make a well in the center and pour in the beaten eggs, yogurt (or soured milk), and the 2 tablespoons vegetable oil. Stir all together.

**3.** Heat 1 teaspoon of vegetable oil on a griddle and drop the batter by the tablespoon when it's hot. Brown approximately 2 minutes each side. Add more oil to the skillet if necessary. Serve hot with maple syrup or molasses.

# DESSERTS

CHAPTER 10

*Thou shalt come to thy grave in a full age, like as a shock of corn cometh in his season.*

—JOB 5: 26

Life was difficult for the early settlers and if it hadn't been for their corn crops, they would not have been able to establish their colonies. It wasn't unusual for them to eat corn three times a day. They ate puffed parched corn with milk and cornmeal porridge ("loblolly" as the British called it) for breakfast; stewed dried corn with squash and beans for dinner, with a little meat thrown in for supper; and then they would enjoy something sweet for a pudding treat. And of course it would be made with ground corn to which they added fruit (blueberries and cranberries were abundant) or molasses and spices for their version of Indian Pudding.

## PUFFED CORN CAKE

PREPARATION: 5 MINUTES          BAKING: 25 MINUTES          YIELD: 4 SERVINGS

*More like a giant baked pancake, this recipe is easy on the cook and ideal for a brunch.*

2 tablespoons melted butter blend or
   margarine
2 large eggs
1 cup low-fat milk
1 tablespoon vanilla extract
2 tablespoons light brown sugar
¾ cup all-purpose flour
¼ cup white cornmeal
½ teaspoon ground cinnamon

**1.** Place a well-greased deep pie dish or small roasting pan in a 450°F. oven and preheat for 10 to 15 minutes.

**2.** Reserving the cinnamon, beat the rest of the ingredients together and pour into the hot baking dish. Sprinkle with the cinnamon.

**3.** Reduce the oven to 400°F. and bake for 25 minutes, until puffed and brown. Serve immediately with assorted jams or honey.

# ORANGE CORN BREAD PUDDING

PREPARATION: 15 MINUTES          BAKING: 45 MINUTES          YIELD: 6–8 SERVINGS

*As a child in England, I ate lots of bread puddings. There were those studded with raisins, dates, and a variety of dried fruits, fresh apples and ginger, marmalade, and, of course, chocolate. No matter the flavor, they were always served with custard sauce. Serve this one with orange sherbet.*

4 cups diced corn bread (half of the corn bread recipe on page 114)
½ cup orange juice
3 large eggs
grated rind of 1 medium orange (Be sure to wash the skin first.)
¼ cup white sugar
1½ cups low-fat milk
¼ cup orange marmalade (or 3 tablespoons brown sugar)

**1.** Preheat the oven to 350°F. and grease a 2-quart baking dish.

**2.** Place the diced corn bread in a large mixing bowl and cover with the orange juice.

**3.** Beat the eggs with the orange rind and white sugar.

**4.** Heat the milk and the marmalade (or brown sugar) in a small saucepan and when it just begins to bubble around the edges, pour into the egg mixture in a thin, slow stream. Stir the mixture constantly.

**5.** Pour this hot custard over the corn bread and stir gently to combine.

**6.** Transfer the batter to the prepared baking dish and set this in a larger pan. Pour boiling water into the outer pan until it comes halfway up the side of the baking dish.

**7.** Carefully place in the oven and bake for 45 minutes, until the top is golden.

# BLUEBERRY SLUMP

*This dish is a bit like a cobbler and is glorious with any kind of fruit. One of my favorite combinations is 2 cups raspberries and 2 cups sliced peaches. I use unsweetened frozen ones when I can't get fresh. You can also try blackberries and apples, but increase the sugar to ¾ cup.*

*Because the fruit is not thickened, there is lots of juice, which makes it unnecessary to serve it with ice cream or a sauce.*

1 cup cornmeal
1 cup all-purpose flour
1 tablespoon baking powder
½ cup brown sugar
⅓ cup butter blend or margarine
½ cup low-fat milk
1 teaspoon vanilla extract (optional)
3–4 cups blueberries
1 tablespoon lemon juice (optional)
½ cup white sugar

**I.** Preheat the oven to 400°F. and grease a 1½–2-quart deep baking dish.

**2.** Combine the cornmeal, flour, baking powder, and brown sugar in a large mixing bowl.

**3.** Rub in the shortening until the mixture resembles coarse breadcrumbs.

**4.** Stir in the milk and vanilla extract until the dry ingredients are just moistened.

**5.** Place the blueberries in the baking dish and sprinkle with the lemon juice and the white sugar.

**6.** Drop the batter on top of the fruit.

**7.** Bake for 35 minutes.

# INDIAN PUDDING

| PREPARATION: 25 MINUTES | BAKING: 1½ HOURS | YIELD: 6 SERVINGS |
| --- | --- | --- |

*There are many versions of this famous colonial dish. Some cooks go heavy on the molasses and butter, so, depending on your taste, you can double the quantities. Also, if you like a rich pudding, use whole milk instead of the low-fat. This dish is delicious served with vanilla ice cream or whipped topping.*

2 cups low-fat milk
⅓ cup molasses
¼ cup yellow cornmeal
¼ cup brown sugar
½ teaspoon ground ginger
½ teaspoon ground cinnamon
½ teaspoon ground nutmeg
3 large eggs
1 cup low-fat milk
2 tablespoons sweet butter blend or
  margarine

**1.** Preheat the oven to 300°F. and grease a 1½-quart baking dish.

**2.** Heat the 2 cups milk and the molasses in a saucepan over low heat and scald until bubbles form around the sides.

**3.** Add the cornmeal, a little at a time, stirring constantly. Cook 10–15 minutes, stirring occasionally until the mixture is thick. Remove from the heat.

**4.** In a small bowl, beat together the sugar, spices, eggs, and milk. Stir into the hot cornmeal mixture with the butter blend. Beat together until thoroughly blended.

**5.** Pour into the dish and bake for 1½ hours.

# APPLE INDIAN PUDDING

*This recipe might seem a little sacrilegious, but it tastes great. It also serves a crowd.*

¼ cup all-purpose or rye flour
½ cup cornmeal
1 teaspoon ground ginger
4 cups low-fat milk
½ cup molasses
2 large eggs, beaten
2 apples (winesap or northern spy), cored
  and chopped with skin

**1.** Preheat the oven to 350°F. and grease a 2-quart baking dish.

**2.** Combine the flour, cornmeal, and ginger in a mixing bowl and beat in 1 cup of cold milk.

**3.** Scald the remaining 3 cups milk in a 2-quart-saucepan and stir into the flour mixture. Return the butter to the saucepan and cook for 10–15 minutes, stirring constantly, until it thickens.

**4.** Remove from the heat and beat in the molasses and the eggs.

**5.** Stir in the chopped apples, pour into the baking dish, and bake for 1½ hours. Serve warm.

# GRITS AND HONEY BARS

PREPARATION: 15 MINUTES   COOKING: 5–6 MINUTES PER BATCH          YIELD: 4 SERVINGS

*These sweet treats are just as good eaten at breakfast or after dinner. You can use cornmeal instead of the grits.*

3½ **cups water**
¾ **cup grits**
2 **tablespoons butter blend or margarine**
¼ **cup warm honey or maple syrup**

**1.** Bring the water to a boil and stir in the grits. When the mixture returns to a boil, reduce the heat and cover the pan.

**2.** Cook slowly for 15 minutes, stirring occasionally. Turn into a greased 9″ × 12″ baking dish, cover with wax paper, and refrigerate for a couple of hours.

**3.** Heat the shortening in a large, heavy skillet.

**4.** Cut the cold grits into 2-inch squares and brown them 2–3 minutes on each side.

**5.** Serve hot with the honey or maple syrup.

# CHOCOLATE PUDDING

COOKING: 15 MINUTES                          YIELD: 4 SERVINGS

*This is not your typical corn pudding, although it qualifies because it uses cornstarch as a thickener.*

1 cup low-fat milk
3 ounces semi-sweet chocolate
⅓ cup sugar
¼ cup cornstarch
¼ cup cold low-fat milk
2 large eggs, separated

**1.** Heat the milk, chocolate, and sugar in a double boiler.

**2.** Stir occasionally until the chocolate has melted.

**3.** Combine the cornstarch with the ¼ cup cold milk and whisk into the hot chocolate mixture until it is smooth.

**4.** Stirring constantly, cook until it thickens, about 5 minutes.

**5.** Beat in the egg yolks. Remove from the heat.

**6.** Beat the egg whites until stiff. Fold into the chocolate mixture.

**7.** Spoon into a 1½–2-quart pudding dish and chill or freeze.

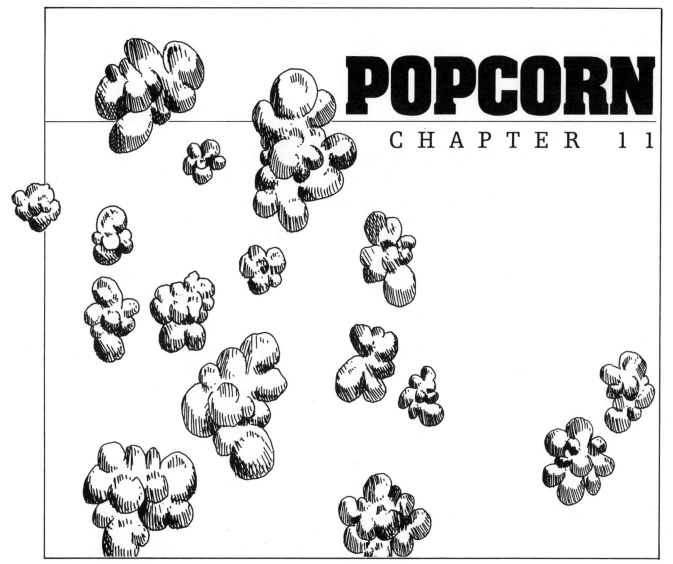

# POPCORN

## CHAPTER 11

*Where the corn is full of kernels*
  *And the colonels full of corn.*
—WILLIAM JAMES LAMPTON (1859–1917)
<u>KENTUCKY</u>

**P**opcorn is a variety unto itself. There is a little extra moisture in the starchy center of each kernel that turns to steam as it heats up. The expanding steam pushes against the hard kernel shell until it bursts open and turns inside out. Most varieties of popcorn expand their kernel size twenty to thirty times, and some of the hybridized "gourmet" popcorn expands forty times. No matter how hard it is to believe, a giant popcorn, called Billion Dollar Baby, has been developed by the plant breeders. This one pops to the size of a grapefruit and measures 14.5 inches around!

Popcorn kernels are available at the supermarket, but if you're looking for something a little exotic like strawberry, calico, or black popcorn, you'll have to grow it yourself or send away for it. It's quite magical to see the black kernels, called Black Jewel, transformed into a snowy white popped corn that literally melts in your mouth. You can order black popcorn from Black Jewel Popcorn, Route 1, St. Francisville, Illinois 62460.

This is a snack that some of us can't get enough of. Granted, there are those who like to douse it heavily with butter and salt, but for those of us who want a healthy low-calorie snack, most of the following recipes will tantalize the tastebuds and still let you tighten your waistband.

---

## PEANUT BUTTER POPCORN

PREPARATION: 5 MINUTES                    YIELD: 4 SERVINGS

---

8 cups popped corn
2 cups raisins
½ cup chunky peanut butter
¼ cup honey or molasses

**1.** Place the hot popped corn in a large serving bowl and mix in the raisins.

**2.** Heat the peanut butter and molasses together and pour over the popcorn.

**3.** Toss to combine and serve.

---

# REGULAR POPPED CORN

PREPARATION: 1 MINUTE          POPPING: 5 MINUTES          YIELD: 8 CUPS (2–4 SERVINGS)

*If you're using an electric popper, or microwave popcorn, follow the manufacturer's directions. The following recipe is for the down-home, skillet-on-top-of-the-stove method.*

*Serve plain or add flavors. If you are adding salt, wait until all the corn has popped. One cup of plain (or flavored only with spices and herbs) popcorn contains only 25 calories.*

*If you have difficulty getting your popcorn to pop, it may have dried out during a long storage. Recondition the kernels by placing them in a 1-quart jar (leave about 2 inches of head room) and adding 1 tablespoon water. Cover the jar and shake a few times. Refrigerate for 2 days until the kernels have absorbed the moisture.*

**2 tablespoons corn oil or butter-flavored oil**

**¼ cup popcorn kernels**

**1.** Warm a heavy skillet over medium-high heat.

**2.** Add the oil so that it just covers the bottom of the skillet.

**3.** After 2 minutes add a few kernels and if one pops, pour in the rest of the corn.

**4.** Cover the skillet immediately and shake to coat all the kernels with oil and to make sure that they will be evenly heated. The cover should not be tight-fitting as this prevents the steam from escaping.

**5.** Shake the pan until the popping slows down. Remove from the heat.

# ITALIAN POPCORN

PREPARATION: 2 MINUTES      YIELD: 2–4 SERVINGS

8 cups popped corn
2 tablespoons butter or margarine
1 teaspoon dried oregano
½ teaspoon dried thyme leaves
¼ cup grated Parmesan cheese
½ teaspoon garlic powder (optional)

**1.** Place the hot popped corn in a large serving dish.

**2.** Melt the butter and combine with the rest of the ingredients.

**3.** Pour over the popcorn and toss to mix. Serve.

# MEXICAN POPCORN

PREPARATION: 2 MINUTES      YIELD: 2–4 SERVINGS

8 cups popped corn
2 tablespoons butter or margarine
½ teaspoon chili powder
½ teaspoon cumin powder
⅛ teaspoon cayenne pepper (optional)
¼ cup grated Parmesan cheese

**1.** Place the hot popped corn in a large serving bowl.

**2.** Melt the butter and combine with the rest of the ingredients and sprinkle over the popcorn.

**3.** Toss to mix. Serve.

# CHIVE BUTTERED POPCORN

PREPARATION: 3 MINUTES                    YIELD: 2–4 SERVINGS

8 cups popped corn
¼ cup chopped chives
2 tablespoons butter blend or butter-flavored oil
1 teaspoon dried basil

**1.** Place the hot popped corn in a large serving bowl and combine with the chopped chives.
**2.** Melt the butter and stir in the basil.
**3.** Drizzle the hot butter over the popped corn and mix together. Serve.

# NUTTY HONEY POPCORN

PREPARATION: 3 MINUTES        BAKING: 15 MINUTES        YIELD: 4 SERVINGS

8 cups popped corn
1 cup roasted peanuts or almonds
¼ cup butter blend or butter-flavored oil
¼ cup honey, molasses, or maple syrup
1 teaspoon ground cinnamon

**1.** Preheat the oven to 250°F.
**2.** Place the hot popped corn in a large roasting pan and combine with the nuts.
**3.** Melt the butter and add the honey and cinnamon.
**4.** Stir into the popcorn and nuts and bake for 15 minutes.
**5.** Cool and serve.

# CORN
# CRAFTS

CHAPTER 12

*"It smells like gangrene starting in a mildewed silo, it tastes like the wrath to come, and when you absorb a deep swig of it you have all the sensations of having swallowed a lighted kerosene lamp. A sudden, violent jolt of it has been known to stop the victim's wrath, snap his suspenders and crack his glass eye right across."*

—DEFINITION OF "CORN LICKER" GIVEN TO THE DISTILLERS' CODE AUTHORITY, N.R.A.

In Colonial times, people devised ways to use just about every part of the corn. Once devoid of kernels, the cobs could be used in numerous ways. The soft inner core was scooped out to make corncob pipes, pill "bottles," and salt shakers. Left whole, cobs were used as handles for knives and other tools. They also became stoppers for jugs and mouseholes. Dried cobs were burned in place of wood, or dipped in oil and set alight they became torches and candles. Women wrapped their hair around shriveled cobs. People used them for back scratchers and as scrubbing brushes for pots and pans. Children played with cobs, making them into dolls and doll houses.

Husks were also put to use. They were wrapped around flat cakes before they were baked in the ashes; they were rolled around ground meat to form sausages. This papery material was transformed into writing paper and cigarette wrappers, which were filled with dried corn-silk tobacco. Shredded husks were stuffed into cushions, pillows, and mattresses, and clumps were bound to sticks and fashioned into brooms to sweep the floors and furniture. Dyed and braided, husks were made into decorative mats and chair covers. Perhaps one of the most enjoyable uses was to turn them into cornhusk dolls and dry flowers.

Cornhusk dolls are still popular today; many people make them to decorate their homes or share a craft with a child. This is particularly true around the Thanksgiving and winter holidays when Pilgrim dolls and Colonial dolls are made to decorate the table, mantelpiece, or Christmas tree. Women dolls are often dressed in little caps and aprons and carry a tiny bucket or broom, while the men wear corn husk pantaloons and shoulder a musket or farm implement. A Christmas tree decorated entirely with cornhusk dolls and dry flowers is a very pretty sight, especially when it is strung with ropes of cranberries and popcorn, following the Colonial tradition.

# CORNHUSK DOLL

*Corn husks can be saved when you're shucking the ears of corn for dinner or you can buy dried husks at a craft shop or a Mexican food store. If you use fresh corn husks, you'll have to dry them first. To do this, remove the husks (or shucks) and place them in a single layer on paper towels. Dry them in the sun for several days, or overnight in a gas oven with heat from only the pilot light. Trim these husks at the top and bottom to give a clean edge and neat appearance.*

**12 corn husks**
**paper towels or an absorbent tea towel**
**2 pipe cleaners**
**1½-inch Styrofoam ball**
**strong thread or fine string**
**2 cotton balls**
**9-inch piece of ribbon**
**fine wool yarn or embroidery silk**
**white or transparent glue**
**2-inch circle of calico or dress fabric**

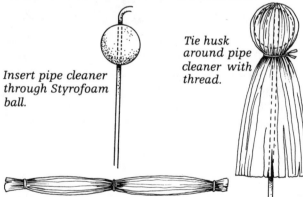

Insert pipe cleaner through Styrofoam ball.

Tie husk around pipe cleaner with thread.

Form arms by rolling pipe cleaner in piece of husk. Tie.

**1.** Soak the corn husks in tepid water for approximately 10 minutes, until they are pliable. Remove and wrap in the paper towels or tea towel until you are ready to use them.

**2.** Insert a pipe cleaner through the center of the Styrofoam ball and hook the end down the back to the "neck." Spread a piece of husk over the ball and smooth it out over the "face." Tie the ends around the pipe cleaner tightly with several layers of thread.

**3.** Take a second pipe cleaner and place it at the bottom edge of a piece of corn husk. Roll the corn husk and the pipe cleaner together to meet the top edge of the husk. Wrap the two ends and the center with thread. Attach the "arms" to the "head," just below the neck and underneath the loose ends of corn husk, by cross-wrapping the arms horizontally around the pipe cleaner body about three times until secure.

*Tie corn husk flaps tightly around waist with thread.*

*Turn doll upside down and gather waist.*

*Smooth gathered skirt down.*

*Pin apron to front of doll. Cross apron straps over back and tie around waist with thead.*

**4.** Place 2 cotton balls under the corn husk to form the chest between the arms. Bring cornhusk flaps down the back and the front and tie tightly around the waist.

**5.** Use the remaining whole pieces of husk for the skirt, reserving one for the apron. Turn the doll so that the head is upside down. With needle and thread, gather each piece of husk and tie around the waist *above* the head, so that the skirt is in actual fact inside out. Repeat until all husks (except the one reserved for the apron) have been gathered around the waist, being sure to overlap the seams with each additional husk. Tie off the final thread and turn the doll the right way up and smooth the skirt down gently so that the gathered waistline is hidden inside.

**6.** Take the final corn husk or dress fabric of your choice and make the apron. Using the narrower end, cut it down the center for about five inches. Carefully pin the apron to the front of the doll so that the cut goes from the waist up towards the shoulders. Take each half up and over the shoulders crossing them over down the back. Tie around the waist tightly with thread (and remove the pin from the front). Use the ribbon as a belt and tie it in a bow at the back. Trim off any excess crossed-over husk showing at the back below the belt.

---

**7.** Glue yarn or embroidery silk to the head to give the appearance of hair and use a felt-tip pen to create eyes, nose, and mouth. Gather the small circle of cloth to fit the head. This will be the dust cap and can be glued in place and trimmed if the brim is too long. Gently position the arms in a desired position and affix a miniature bucket, broom, or bouquet of flowers.

**8.** If the cornhusk doll is to be hung on a door or Christmas tree, tie a loop of thread around the neck.

## CORNCOB DOLL

*You can make a doll from the leftover cobs after a corn-on-the-cob dinner, or you can buy an ear specifically for the purpose of making a doll. To dry a corncob, remove the kernels and dry naturally in a warm place for 3–4 weeks, or place in a gas stove and dry for 3–4 days with the aid of only the pilot light.*

white or transparent glue
dried corn silk or strands of fine wool
   yarn
1 dried corncob
12 whole cloves
12-inch-square piece of calico or light
   dress fabric
9-inch piece of ribbon
pipe cleaner (optional)

**I.** Glue the dried corn silk or strands of wool to the top of the dried cob to resemble hair.

**2.** Glue the cloves on to the "face" to represent eyes, nose, and a mouth.

**3.** Make a dress by folding the fabric diagonally and cutting a 1–1½-inch slit in the middle of the fold. Push the doll into the dress until it reaches the "neck."

**4.** Tie the ribbon to make a belt around the "waist."

**5.** If desired, use a long pipe cleaner wrapped around the cob to make arms.

# BASIC CORNHUSK FLOWER

*Keep the corn husk intact by carefully peeling it back and snapping the corn cob off the stem.*

**1.** Soak this in water for 5 minutes. Form it into its original cob shape and cut it across approximately 3½ inches above the stem.

**2.** Slightly separate each layer of husk to form petals.

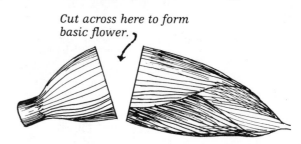

*Cut across here to form basic flower.*

# DAISY

1 Basic Cornhusk Flower, above
1 single cornhusk leaf
white or clear glue
thread

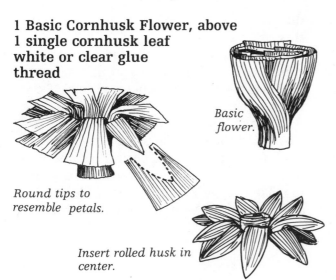

*Basic flower.*

*Round tips to resemble petals.*

*Insert rolled husk in center.*

**1.** Dip the basic flower in water for 2 minutes. Pull the "petals" of the basic flower all the way down into a flat circle. Cut the petals into ½-inch strips and round the tips to resemble daisy petals.

**2.** Cut ½-inch strips of husk from along the width and roll them together until you have enough to fill the center. Glue the roll together and then fill the hole with glue and set the roll within. Hold the flower, pressing on the center, until the glue has set. (A glue gun would speed up this process.)

**3.** Make a longer stem following the directions for the Chrysanthemum.

# CHRYSANTHEMUM

1 Basic Cornhusk Flower, page 165
1 single cornhusk leaf
white or clear glue
thread

*Basic flower.*

*Roll strip of husk to form center.*

*Snip roll into strips.*

**1.** If the basic flower has dried out, dip it in water for 2 minutes. Cut the petals in thin strips almost down to the stem and separate them.

**2.** Fill the hole in the middle of the flower by rolling 2½-inch-wide strips of husks (cut along the width, not the length) together until there are enough to fill the center. Glue together at the bottom and pour some glue into the base of the center of the chrysanthemum. Press the roll in place and hold for a minute or two until the glue sets. (A glue gun would speed up this process.)

**3.** When the glue has dried completely, carefully snip the center roll into strips to resemble chrysanthemum petals. If desired, cut points at the tip of each strip.

**4.** Make longer stem by attaching a pipe cleaner to the existing stem with thread. Spread some glue on a corn husk and wrap it around the pipe cleaner and corn stem to make it appear as one. Cut out leaves and attach them to the stem with glue or thread.

# INDEX

Boldface numbers such as **55** indicate that illustrations appear on that page.

---